Monitoring Genetically Manipulated Microorganisms in the Environment

Wiley Biotechnology Series

Series Editors: Professor J. A. Bryant, *Department of Biological Sciences, Exeter University, UK*, and Professor J. F. Kennedy, *Department of Chemistry, University of Birmingham, UK*.

This series is designed to give undergraduates and practising scientists access to the many related disciplines in this fast developing area. It provides understanding both of the basic principles and of the industrial applications of biotechnology. By covering individual subjects in separate volumes a thorough and straightforward introduction to each field is provided for people of differing backgrounds.

Published Titles

Biotechnology: The Biological Principles   M. D. Trevan, S. Boffey, K. H. Goulding and P. Stanbury

Fermentation Kinetics and Modelling   C. G. Sinclair and B. Kristiansen (Ed. J. D. Bu'lock)

Enzyme Technology   P. Gacesa and J. Hubble

Animal Cell Technology: Principles and Products   M. Butler

Fermentation Biotechnology: Principles, Processes and Products   O. P. Ward

Genetic Transformation in Plants   *R. Walden*

Plant Biotechnology and Agriculture   K. Lindsey and M. Jones

Biosensors   E. Hall

Biotechnology of Biomass Conversion   M. Wayman and S. R. Parekh

Biotechnology in the Food Industry   M. P. Tombs

Bioelectronics   S. Bone and B. Zaba

An Introduction to Fungal Biotechnology   M. Wainwright

Biosynthesis of the Major Crop Products   P. John

Monitoring Genetically Manipulated Microorganisms in the Environment

Edited by

Clive Edwards

Department of Genetics and Microbiology
University of Liverpool, UK

JOHN WILEY & SONS
Chichester · New York · Brisbane · Toronto · Singapore

Other Wiley Editorial Offices

John Wiley & Sons, Inc., 605 Third Avenue,
New York, NY 10158–0012, USA

Jacaranda Wiley Ltd, G.P.O. Box 859, Brisbane,
Queensland 4001, Australia

John Wiley & Sons (Canada) Ltd, 22 Worcester Road,
Rexdale, Ontario M9W 1L1, Canada

John Wiley & Sons (SEA) Pte Ltd, 37 Jalan Pemimpin #05–04,
Block B, Union Industrial Building, Singapore 2057

Library of Congress Cataloging-in-Publication Data

Monitoring genetically manipulated microorganisms in the environment/
edited by Clive Edwards.
    p.   cm.—(Wiley Biotechnology series)
    Includes bibliographical references and index.
    ISBN 0–471–93795–9
    1. Recombinant microorganisms—Environmental aspects.
    2. Environmental monitoring.   I. Edwards, Clive, 1949–.
    II. Series.
    QR100. M65 1993
    660′.65—dc20   92–30174
                                                                CIP

British Library Cataloguing in Publication Data

A catalogue record for this book is available
from the British Library

ISBN 0 471 93795 9

Typeset in 10/11½ pt Times by Cambridge Composing (U.K.) Ltd, Cambridge
Printed and bound in Great Britain by Biddles Ltd, Guildford, Surrey

Contents

Contributors

Dr W. Amner   Dept of Genetics & Microbiology,
                         University of Liverpool,
                         PO Box 147,
                         Liverpool L69 3BX

Dr N. Cresswell   Dept of Biological Sciences,
                         University of Warwick,
                         Coventry CV4 7AL

Dr C. Edwards   Dept of Genetics & Microbiology,
                         University of Liverpool,
                         PO Box 147,
                         Liverpool L69 3BX

Dr P. R. Herron   Dept of Biological Sciences,
                         University of Warwick,
                         Coventry CV4 7AL

Dr D. W. Hopkins   Dept of Biological Sciences,
                         University of Dundee,
                         Dundee DD1 4HN

Dr A. J. McCarthy   Dept of Genetices & Microbiology,
                         University of Liverpool,
                         PO Box 147,
                         Liverpool L69 3BX

Dr J. A. W. Morgan   Freshwater Biological Association,
                         Windermere Laboratory,
                         Ambleside,
                         Cumbria LA22 0LP

Dr A. G. O'Donnell   Dept of Agricultural and Environmental Science,
University of Newcastle upon Tyne,
Newcastle upon Tyne NE1 7RU

Dr R. W. Pickup   Institute of Freshwater Ecology,
Windermere Laboratory,
Ambleside,
Cumbria LA22 0LP

Dr J. I. Prosser   Dept of Molecular & Cell Biology,
University of Aberdeen,
Marischal College,
Aberdeen, AB9 1AS

Dr V. A. Saunders   School of Biomolecular Sciences,
Liverpool John Moores University,
Rodney House,
70 Mount Pleasant,
Liverpool L3 5UX

Dr J. R. Saunders   Dept of Genetics & Microbiology,
University of Liverpool,
PO Box 147,
Liverpool L69 3BX

Dr E. M. H. Wellington   Dept of Biological Sciences,
University of Warwick,
Coventry CV4 7AL

Dr C. Winstanley   Dept of Genetics & Microbiology,
University of Liverpool,
PO Box 147,
Liverpool L69 3BXP

Chapter 1

The Significance of In Situ *Activity on the Efficiency of Monitoring Methods*

C. Edwards

Department of Genetics & Microbiology, University of Liverpool

INTRODUCTION

Numerous articles have appeared that discuss the fate and possible consequences of releasing genetically engineered microorganisms (GEMs) into
natural environments (9,15,21). Rarely have these issues been addressed
from the physiological viewpoint of whether a bacterium would be able to
capture sufficient energy substrate for the synthesis of enough ATP to allow
for growth. Most often, bacteria in natural environments grow extremely
slowly with long cell cycle times, which in turn increase the proportion of
ATP that has to be directed towards the maintenance energy (Me) requirement (53). The higher the Me in terms of ATP, the less energy that is
available for growth purposes. As pointed out by Chesbro *et al.* (8), slow
growth imposes severe stress on cells. Gaps from which chromosome
replication is absent develop and the proportions of cell cycle stages change
so that the cell growth cycle is completely unbalanced. Adapting to the
starved state results in populations with an average chemical and molecular

Monitoring Genetically Manipulated Microorganisms in the Environment. Edited by C. Edwards
Published 1993 John Wiley & Sons Ltd. © 1993 C. Edwards

composition that is markedly different from exponentially-growing, nutrient-sufficient cultures (8). The following questions must therefore be posed when considering the fate and survival of GEMs in natural environments:

1. How does nutrient status affect survival?
2. How does nutrient status affect expression of cloned gene products?
3. Will GEMs be more or less stable than indigenous species?
4. How may interaction of GEMs with indigenous species be monitored?
5. Do natural environments affect bacterial phenotype?
6. What techniques are available that can identify GEMs when present in heterogeneous populations as well as measure their activity *in situ* irrespective of the stresses imposed by natural environments?

Obviously, in some specialized cases these questions do not pose any difficulties. Where a GEM is used in pure culture in an environment where nutrients are in excess, predictions can be made more easily, very much on the basis of observations that have already been made in laboratory culture. Such specialized applications might include methane production from waste materials or biodegradation of natural substrates such as lignocellulosics. In natural environments, *nutrient limitation* will be the norm (16) and for heterotrophic bacteria this will pose considerable problems for survival, assuming of course that this is desirable.

As we shall see, nutrient limitation affects heterotrophic bacteria in a number of ways, particularly with respect to proliferation and activity. Some specialized species such as *Bacillus* and actinomycetes have the option of differentiating to a dormant form as metabolically inactive spores. What has become increasingly unexpected, is the way non-differentiating heterotrophs can survive for long periods in nutrient-limited habitats. This results in very profound changes in both structure and activity and under conditions of starvation it is difficult to see how GEMs are ever going to express activities of cloned gene products at sufficient levels to mediate an efficient environmental function.

The focusing of attention on the fate and consequences of the release of GEMs into natural environments has had an unexpected but highly significant bonus. Molecular techniques have been directed not only at the bacterium of interest but also at indigenous species within the habitat of interest. This has led to a proliferation of interest in all aspects of microbial ecology and led to the applications and development of new techniques for the study of community structure. As we shall see later, many of these methods rely on the application of sophisticated, and in many cases expensive, equipment not previously used in studies of microbial ecology. They also allow sensitive measurements of the consequences of release of GEMs into stable, indigenous microbial communities.

These then form the topics for discussion in this chapter—effects of nutrient limitation pertinent to survival of GEMs and description of emerging technologies that can be used either directly to identify them in heterogeneous populations or that allow sophisticated, and in some cases rapid, methods, for studying their effects on community structure and/or activity.

ENVIRONMENTAL EFFECTS

Environment–cell interactions

Figure 1.1 shows a generalized scheme of the ways in which microorganisms are known to deal with an *adequate* substrate supply. When the substrate is polymeric, hydrolytic enzymes are secreted from the cell to break the polymer down to more easily assimilated monomers. The fate of most carbon substrates is oxidation, resulting in the production of biomass together with end-products of metabolism. For some substrates or in certain species oxidation may be inefficient so that a large proportion of ATP produced is either wasted or directed to non-growth purposes. Highly efficient coupling of substrate catabolic reactions to those of anabolism results in high biomass. Products are many and varied and include end-products of metabolism such as organic acids or secondary metabolites (e.g. antibiotics) or enzymes. For specialized substrates no net growth-linked oxidation occurs, but these may be transformed to a derivative by special enzymes produced by the organism. Under certain conditions, particularly of nitrogen limitation, carbon substrates may be converted into polymers such as poly-hydroxybutyrate (PHB). In some species nutrients can trigger the pathogenic state, manifest by synthesis of virulence factors and/or toxins. Many of the above processes are well understood for laboratory-grown cultures. The deliberate or accidental release of GEMs into natural environments makes it imperative that we study how bacteria respond to and survive in situations where the substrate input and output responses described in Fig. 1.1 will be grossly modified in nutrient-limited ecosystems.

In natural environments, bacteria find themselves under fluctuating chemical and physical conditions. They have evolved strategies to cope with rapid change and most free-living species display a high degree of metabolic versatility. Some of the major responses evoked by changing nutrient status

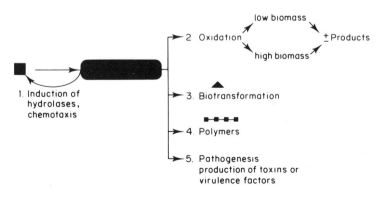

Fig. 1.1 Schematic presentation of some of the responses of a bacterial cell challenged with a substrate (■).

are shown in Table 1.1. Because nutrient-limitation in natural environments is a condition that has probably existed for millions of years, then selective pressure has always existed for the evolution of species able to respond rapidly or survive periods of 'feast/famine'. It is hardly surprising that a wide variety of bacterial species have developed both common and specific methods for coping with nutrient-limitation. The response to starvation must be such that an organism can compete effectively for any available nutrients and at the same time maintain as high a growth rate as possible (50). This allows us to predict that the best adapted species will be able to modulate gene expression in a sophisticated way such that genes are switched on or off in response to environmental signals (50). The usual situation in natural environments is therefore one of *nutrient limitation*. The response of a non-differentiating heterotroph will result in inhibition of all outputs described in Fig. 1.1 other than possibly toxin production (see Ref. 31). However, this will not be a static response and the end-result will not be an inactive form. Such a conclusion is of obvious importance to the survival of GEMs in natural environments.

Table 1.1 Some of the major changes that can occur when bacteria respond to starvation (taken from Refs. 24, 30, 34, 41 and 50).

Growth stops but cell division continues to form non-motile, fimbriated ultramicrocells

cAMP levels high, ppGpp (tetraphosphorylated guanine) levels switch on the stringent response

Maintenance energy requirement increases up to 50–70% of total energy flux

Energy reserves (e.g. PHB) exhausted

Toxin secretion by some pathogens

Development of resistance to UV light, heat and autolysis

Enlargement of the periplasm

Half-life of mRNA increased

Starvation inducible (*sti*) proteins synthesized, some are long-lived proteins, others appear only transiently

rRNA content decreases, e.g. to 10–20% of original levels after starvation of *Vibrio* sp. for 15 days

Some species become non-culturable, i.e. cells are viable but cannot be recovered by conventional plating techniques

---

Adaptation to low nutrient environments

The effects of nutrient depletion on non-differentiating bacteria have received considerable attention over the last decade. A number of important findings have resulted from this work, the one most germane to GEMs being that heterotrophic bacteria can develop strategies that ensure their survival for long periods in both aquatic and soil environments. A well-documented response is the development of uptake systems that have low affinity constants; these ensure efficient capturing of essential molecules such as sugars, amino acids and ions, even under low nutrient concentrations (16).

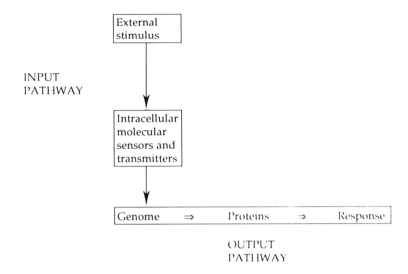

Fig. 1.2   Generalized diagram of stimulus-response pathways in bacteria.

The response to starvation is thought to be typical of the many stimulus-response networks that have now been identified in bacteria. A generalized scheme for such networks is shown in Fig. 1.2. These networks involve an external, environmental stimulus which affects a specific cellular target. This generates a molecular signal to a regulator molecule which in turn can modulate the activity of cellular molecules to cope with the stimulus. The most usual situation is a transcriptional regulator that initiates transcription of genes which encode for proteins that will deal or respond to the external stimulus (7,34). These stimulus-response networks therefore transmit information external to the cell and invoke a suitable response. Examples of effector stimuli include the SOS response to UV light damage to DNA and the heat-shock response in *E. coli* (7, 34). Together they constitute global networks that respond to external stresses. It is now accepted that nutrient starvation also results in a stimulus-response network in many bacteria. This work has predominantly centred on species of *Vibrio*, *Salmonella* and *Escherichia coli* and has led to an understanding of the molecular responses to starvation which in turn are important for understanding the fate and survival of GEMs in natural environments.

Possible responses of bacterial species to starvation are listed in Table 1.1. These are not necessarily universal to all heterotrophs but serve as a benchmark for possible changes that may result when GEMs are released into natural environments. Some of these also have general implications for the methods developed for detection and identification.

One of the major effects of starvation is the possible generation of cells altered in morphology and biochemistry. The ultramicrocells of *Vibrio*, for example, are not only different in size but also in appearance (34). Microscopy may therefore fail to detect these new forms. Differences in biochemistry, particularly in surface layers, can change the antigenicity of the cell. Methods based on immunological identification (e.g. fluorescent antibodies) may not work if the antisera were raised against antigen profiles of well-fed laboratory cultures. Environmental signals are certainly known to control the expression of virulence determinants in bacteria (31). Another ramification of these observations is that a GEM not normally pathogenic in laboratory culture may express hidden virulence determinants in the environment. Factors such as iron availability (*Corynebacterium diphtheriae*, *Escherichia coli*, *Pseudomonas aeruginosa*), CO_2 levels (*Bacillus anthracis*, *Vibrio cholerae*), temperature (*Bordetella pertussis*, *Shigella* sp.), carbon source (*E. coli*), amino acids (*V. cholerae*) have all been identified as modulators of the expression of virulence determinants (31). Another feature of starving cells is the gross changes that occur in macromolecular composition (Table 1.1). For many detection systems this means less sensitivity.

A good example here is the decrease in rRNA content that ensues on starvation. This lowers the sensitivity and applicability of 16S rRNA oligonucleotide probes so significantly as to make detection extremely difficult. Kramer and Singleton (24) showed that relative hybridization of a 16S rRNA probe against *Vibrio* rRNA decreased to between 10 and 26% after 15 days starvation.

Another major change occurs in the proteins of cells. Many genes expressed during active vegetative growth will be repressed during conditions of nutrient-limitation. Profiling methods that rely on 'fingerprint'-type analysis of macromolecular composition, e.g. lipids, proteins, will only work if the databank of collated profiles of known origin is derived from cells which have been subjected to starvation. The differential gene expression that occurs when cells are placed into a nutrient-depleted environment has further consequences for methods designed to track GEMs. If as in the heat-shock response the genes encoding *sti* proteins have unique promoters that rely on alternate sigma factor-promoted modulation of RNA polymerase, then cloned genes, including any marker or reporter genes, may not be expressed in such environments. This situation is obviously reassuring when GEMs are released into ecosystems accidentally. However, it will be a major consideration if the cloned gene product has been designed to carry out specific functions, e.g. bioremediation purposes. One way to circumvent this problem advanced by Matin (30) is to ensure that the recombinant DNA is under the control of the starvation promoters.

Finally there is now mounting evidence that certain bacterial species are able to adopt a viable but non-culturable form within natural environments. The formation of ultramicrocells (see Table 1.1) by reductive cell division is thought to be the reason for this and is a feature reported for *Aerobacter*, *Flavobacterium*, *Cytophaga* and *Vibrio* spp. Ultramicrocells are metabolically active, albeit at a much reduced level, but cannot be cultured by conventional plating methods (41). They may therefore constitute a reservoir, unknown in magnitude, of live cells that are essentially invisible to conventional recovery methods. Recently McKay (28) has briefly reviewed viability and the phenomenon of non-culturability of pathogenic aquatic bacteria. Non-culturability has been particularly ascribed to those pathogenic species capable of growth in warm-blooded hosts and this condition may be promoted or enhanced by sub-lethal environmental stresses such as sub-optimal temperatures, salinity or pollutants. Species listed by McKay as demonstrably viable but non-culturable include *Legionella pneumophila*, *Campylobacter* spp., *Vibrio* spp., *Escherichia coli*, *Salmonella enteritidis* and *Shigella* spp.

Another factor worthy of mention here is that bacterial growth and survival in natural environments can be significantly affected by attachment to particles or surfaces. It has been found that in a natural population of microorganisms, attached bacteria are often more active than free cells (27). Examples include the observation that glucose mineralization in estuaries is predominantly carried out by adhered bacteria; in soils, degradative activity was essentially carried out by attached organisms; in a marsh estuary, almost all detectable and respiring bacteria were associated with particles (see Ref. 27 for review). However, although attachment can change the properties and activities of microorganisms, it is hard to predict what changes will occur after attachment. One consequence will be the formation of a biofilm whereby cells become firmly attached to an inert support. Biofilms pose considerable problems for identifying component species and assessing the numbers of viable cells present and constitute an area meriting further work for unmasking biofilm community structure and interactions—particularly with respect to GEMs.

In conclusion, release of GEMs into natural environments can have profound effects on phenotypic expression. Properties of cells in natural environments will differ greatly to those in the laboratory. These changes result in altered efficiency of detection, and the phenomenon of viable but non-culturable forms displayed by some species makes their identification and recovery impossible by conventional methods. In the next section of this chapter, novel, mainly biophysical, methods are described for monitoring activity and for specific detection and recovery of GEMs in natural environments. These methods fall into two general groups based on *identification* or *activity*, although many techniques can be used for concomitant assessment of both. Some of these methods are difficult to apply as tracking or identification methods for a single species; however, they are eminently suited to monitoring gross changes in community structure. This is crucial if the impact and consequences of release of GEMs into open environments are to be monitored.

IDENTIFICATION METHODS

General microscopic techniques

Microscopic techniques applied to studies of bacterial ecology have benefited from a number of technological and biological developments. Technological advances include the use of computer software for process control and data handling, whilst biological ones centre on the ability to prepare fluorescent monoclonal antibody or nucleic acid probes that have such a high degree of specificity as to make possible strain-specific identification protocols.

A good example is the use of fluorescent oligonucleotide probes directed at 16S rRNA sequences. Ribosomal RNA is well suited as a target for identification of GEMs: it is universal, comprises a mixture of conserved and variable regions, is stable and coded for by genes that are often present as multiple copies. Sequences of varying specificity are available that are characteristic for strains, species or genera. In addition universal probes for eubacterial species or archaebacterial ones are available which makes these oligonucleotides exquisite molecules for investigating community structure (58).

Fluorescently labelled 16S rRNA oligonucleotides have been used for microscopic identification of single cells (1, 11) as well as for analysis of community structure (2). Amann et al. (3) recently studied a population of sulphidogenic biofilms using fluorescence microscopy and probes complementary to a region of 16S rRNA common to sulphate-reducing bacteria. Epifluorescence microscopy was used to visualize specific sulphate-reducing populations within biofilms that had been labelled with the appropriate fluorescent oligonucleotide probe. These workers concluded that 'combined use of PCR amplification, comparative sequencing and oligonucleotide probe hybridization offers the basis for a systematic dissection of biofilm microbial community structure' (3) [PCR = polymerase chain reaction].

Image analysis

A problem with conventional microscopy is the time taken to view and enumerate stained cells. This, coupled to a degree of subjectivity required for enumeration and sizing, has led to the development of automated methods for deriving quantitative data from images. A basic set-up for applications of image analysis is shown in Fig. 1.3. Two features, which stem

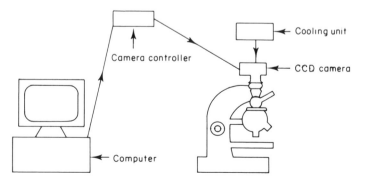

Fig. 1.3 Features of an image analysis system.

from technological developments in the last decade, make this a powerful technique. First, a dedicated microcomputer is used to control camera operation and process the data obtained. Second is the availability of cooled charge-coupled device (CCD) cameras which can be attached to a microscope via the objective. In some systems a video camera is employed, but a CCD camera is better suited to bacterial sizes (54). CCD cameras, originally used in astronomy, are cooled and scan slowly, which means that random electronic and thermal noise is eliminated. Exposure times may be varied to suit the intensity of emitted light—typically seconds for fluorochromes and up to 10 to 30 min for luminescing bacteria.

Studies on marine bacteria revealed that conventional fluorescence microscopy failed to detect the bacterial minicells that dominated oceanic samples (55). Development of an image analyser system resulted in better detection, enumeration and sizing of planktonic bacteria (47). Such a development has obvious implications for detecting ultramicrocells present in natural environments (see above; 15, 41). A theoretical and statistical account for the evaluation of image analysis for accurately sizing fluorescent cells has been described by Sieracki *et al*. (48).

A more recent development and application of image analysis has exploited progress in the molecular genetics of reporter genes. The luciferase system that employs cloned *lux* genes (44) results in emission of bioluminescence by the recombinant host. It therefore functions as a model reporter gene and gives some indication of viability since dead cells do not emit light. A number of factors predispose to the use of *lux* genes and these are dealt with elsewhere in this book. However, coupled to image analysis systems employing CCD cameras, *lux* genes show some potential for tracking GEMs in natural environments. Shaw *et al*. (45) have shown the potential of this approach in a study of *Xanthomonas campestris* genetically engineered to bioluminesce by insertion of *lux* genes. Bioluminescent *X. campestris* could be detected by the CCD camera system up to 6 weeks after inoculation into soil. Bioluminescent enumeration compared well with colony-forming units (cfu) determined by conventional plating methods. Bioluminescent *X. campestris* could also be visualized on plant leaves and in the rhizosphere. Thus, bioluminescence serves as a useful model reporter function for tracking and detecting a GEM in natural environments as well as giving an indication of the numbers of viable cells present and having the potential for spatial analysis.

Confocal microscopy

Confocal microscopy is a new miroscopical technique which has many advantages over conventional microscopy (46). Its use in microbiology is currently limited but its potential in this area has been described by White (57). In the confocal mode, light from the in-focus area is allowed to pass to the detector; out-of-focus information is eliminated. This reduction of out-of-focus blur allows non-invasive optical sectioning by confocal scanning microscopy; when allied to computing facilities it allows imaging and three-dimensional tomography of stained biological specimens.

Laser light or a bright arc lamp is sharply focused on the specimen—this is the illumination probe. Light emerging from the focused specimen is then collected and brought to a second focus equivalent to or 'confocal' with the light source. Both illumination and detection probes contribute equally to the performance of a confocal microscope. CCD cameras are used to improve fluorescence imaging. This methodology, like that of image analysis, could be used to analyse community structure, as long as suitable fluorescent probes are available to detect a target such as a GEM. One possible area where this technique might be useful would be in analysis of spatial structures and distributions of GEMs in complex biofilms.

'Fingerprint' methods

This section deals with what could loosely be described as the 'stamp collecting' approach for detection of GEMs in environments as well as for sophisticated approaches to analysis of community structure. Whereas the fluorescent antibody and nucleic acid probes have exquisite specificity the methods described here have less resolution. What brings them to the fore are developments in computing and data handling as well as what they can tell us about microbial composition under different environmental constraints. The principles of these approaches are first described followed by the way in which the massive amounts of information they generate can be interpreted and possibly used to provide at least an automated method for detection of a GEM. These methods tell us little concerning *viability* of a given target bacterium but do have much promise in quantifying changes in population dynamics within natural environments.

Whole cell analysis

Fourier transform infrared (FT-IR) spectroscopy   The potential of infrared (IR) spectroscopy for differentiating bacteria goes back as far as the late 1950s (33, 42). IR spectra represent the total chemical composition of cellular components such as proteins, lipids, cell wall and nucleic acids represented as vibrational and rotational motions of their atoms. Specific IR absorptions can be assigned to particular covalent bonds (32). The IR portion of the spectrum is at low energy and any changes or analysis of this region applied to environmental samples has proved difficult. Fourier transform IR instrumentation has overcome the problems of low transmittance in heterogeneous samples. Advances in computing have meant that spectral subtractions can also be performed to gate out unwanted signals.

FT-IR is a non-destructive technique that has been used to produce spectra of bacteria and categorize these as a database, which can then be used as a reference for identifying unknown bacteria in environmental samples (32).

FT-IR spectra are fingerprint-like patterns which are highly reproducible and typical for different bacteria (19). Thus the method relies on features within an FT-IR spectrum that are unique to a particular bacterial genus, species or even strain. Helm *et al.* (19) developed a computer-aided, fast-working procedure suitable for the identification of bacteria. They created a spectrum library for 97 different bacterial strains that comprised Gram-negative and Gram-positive species; 72 of these were then randomly selected and repetitively measured and the data used to challenge the library. They were able to obtain an overall quotient of correct matches of 83.3% at the strain level. Most of the mis-identifications resulted for Gram-negative species.

In a later study, Helm *et al.* (20) assessed the ability of FT-IR to allocate bacteria to the correct genus and species, to group bacteria according to one single biological property and to detect subtle differences within a population of very closely related organisms. The database used contained 139 bacterial reference spectra. These workers concluded that FT-IR patterns could be used to type bacteria, provide data enabling correct classifications to be made and that it could be used as an easy and safe method for rapid identification of clinical isolates.

In the context of identifying GEMs, FT-IR shows some promise providing that suitable databases are available. An additional possibility would be to insert or manipulate the dominant covalent bonds, which this technique measures, in a characteristic way so that the GEM of interest had a unique molecular fingerprint. Modified bonds have been shown to be detectable by FT-IR (32).

Pyrolysis　　In this technique a sample, for example containing the bacteria of interest, is heated at a controlled high temperature in a non-oxidizing environment. Heat cleavage of covalent bonds occurs yielding a mixture of low-molecular mass volatile compounds, termed the *pyrolysate*. The pyrolysate can then be analysed by gas–liquid chromatography or, more appropriately, mass spectrometry to yield a spectrum that is a profile or chemical signature characteristic for the sample. The data from microorganisms are complex but classification, identification and typing of microorganisms may be possible (29). Commercial instruments are now available with advanced software. The method relies on generating a databank of profiles for single species and challenging it with unknown profiles. Sophisticated computer software enhances potential applications of this and other fingerprinting techniques.

Macromolecular profiles

The pioneering work of White and co-workers has led to a battery of methods that target macromolecular pools for analysis. Basic approaches to these methods have been outlined by White (56). These include analysis of specific polymers (e.g. β-hydroxybutyrate), typing of triglycerides and wax esters by thin-layer chromatography, wall components and in particular lipid profiles, many of which can be diagnostic for a microbial species. These methods are particularly useful for monitoring community structure to a highly definitive degree. For example, relative proportions of eukaryotes:prokaryotes can be determined from relative proportions of key lipids; phospholipids predominate in bacteria, polyenoic fatty acids with more than 20 carbons are characteristic of eukaryotes. Genera of the Archaea may be identified and quantified on the basis of their unique glycerol ether-linked lipids. Plasmalogens are unique lipids for anaerobic bacteria; likewise many anaerobic fermenters possess unique phosphosphingolipids. Recently Rajendran *et al.* (40) measured the distribution and amounts of phospholipid ester-linked fatty acids (PLFA) in sediments to quantify microbial biomass, community structure and nutritional status. PLFA are particularly useful because they are essential components of bacteria (but *not* the Archaea) and have rapid turnover in sediments. Their analytical quantification can therefore also provide a good measure of the viable cell biomass and it is possible to relate their amounts to bacterial numbers as well as discriminate the proportion of aerobic:anaerobic species (40). Fatty acid profiles of 773 strains representing 25 taxa of plant pathogenic and related saprophytic bacteria were compared with commercially available broad-spectrum libraries—the Microbial Identification System (Microbial ID Inc., Newark, USA). The accuracy at the specific level was often 100% although some closely related species could be significantly less than this. Identifications were routinely made within 48 h and the necessity for standardized cultural conditions was stressed. All in all the protocol offers a valuable rapid identification method for bacteria (51). It should be noted that this type of identification technique relies on the ability to culture the cells prior to analysis.

Another fingerprinting method that has been advanced is that of electrophoretic polymorphism of enzymes or total cell proteins. It appears that even different strains of the same species exhibit subtle differences in the banding pattern of their proteins, which are detected after electrophoretic analysis. Thus a characteristic protein migration profile may be obtained for a given microorganism and stored in a databank. Future isolation and analysis may be identified by feeding data into the databank. This method and approach has been used to analyse strains and species of *Yersinia* (17, 36). Advances in laser densitometry for electrophoretogram analysis coupled to sophisticated computer software herald increasing potential for this method.

Finally, DNA fingerprinting has also been used for identification and analysis of unknown species. This method relies on generating a unique banding pattern of DNA fragments after treatment of DNA with two or more restriction enzymes. The major problems centre on the choice of restriction enzymes and data handling sensitive enough to allow identification. This method has been well reviewed by Forbes *et al.* (14).

The methods described above have a number of potential advantages and also disadvantages. Providing a GEM has a unique cellular or macromolecular chemical signature, then a profiling method can be extremely sensitive.

Of course, one way forward would be to deliberately genetically engineer such a signature into a GEM planned for release into heterogeneous communities. An additional and potentially powerful application of these techniques is to analyse the effects of GEMs on community structure and stability. For example, such an approach could be used to ensure that key microbial groups are not displaced by the GEM. Problems arise with these methodologies regarding reproducibility and physiological status of the cells. At the beginning of this chapter, ways in which bacteria are able to change their biological and chemical properties in the environment in response to nutritional status were discussed (see also Table 1.1). This would no doubt change cellular or macromolecular profiles with the result that they were no longer recognized by a databank. Stemming from this is the sophisticated data handling that is essential for rigorous identification of target species by profiling methods. Many of these methods have only come to the fore because of advances in computing. Further developments in data handling and manipulation may yet make profiling techniques powerful methods in microbial ecology.

One such emerging area is that of computer neural networks which operate, like the brain, in parallel and result in properties not exhibited by conventional computing systems (6). Neural networks have an input layer that receives information, a hidden layer that networks and processes the input signals before passing on instructions to output signals. So far, neural networks have not been widely used but may find applications in microbial ecology, particularly in the profiling techniques discussed above. They are capable of 'learning' from repeated input of data so that when challenged with an unknown data set can recognize it, providing of course that the network has encountered it before. Neural networks have been cited as of particular application to microbial ecology where data are too complex to model by standard numerical methods (6).

ACTIVITY

Methods for monitoring metabolic activity of GEMs in natural environments are limited; indeed measurements of total biomass activity are often the subject of controversy and debate. Recently, this has been fuelled by the recognition that some part of the biomass may be viable but non-culturable and assigning activity to any one microbial group, e.g. nitrifiers or methanogens, will only result in an average value. This is because some species or cells will be highly active, others less so, whilst a proportion may well be inactive but still viable. However, activity measurements are included here as a means of understanding any population perturbations that may result in the biosphere when GEMs are introduced. These approaches are crucial to ensure that key species, such as nitrifiers, are either not displaced or are active to an unfavourable or damaging extent.

One approach that has been adopted is to measure the partitioning and fate of radioisotopes. This has been particularly useful for monitoring activities associated with key steps of the nitrogen cycle. Introduction and subsequent monitoring of the fate of ^{15}N-labelled NO_3^- can be used to investigate nitrification, denitrification and dissimilatory reduction of NH_4^+. This approach has been adopted and described by Binnerup *et al.* (4), who used it to study nitrogen and oxygen transformations in estuarine sediments.

Another technique has involved the use of quadrupole membrane-inlet mass spectrometry (QMS) for measuring gross population activities using gaseous exchange reactions. Quadrupole mass spectrometers ionize molecules by means of an ion source, which is usually a beam of electrons. The ionized molecules are focused onto an analyser that separates them according to values of mass charge ratios (*m/z*) before detection. So, for example, methane has an *m/z* value of 16, nitrogen oxide, 30. Providing *m/z* values are sufficiently different, a variety of gases associated with key processes may be monitored. The potential advantages of mass spectrometry have been proposed (25) to be:

1. Non-invasive and applicable to pure or mixed cultures.
2. Continuous measurements can be made over a considerable range of time—in the order of seconds to months.
3. A number of different gases can be measured simultaneously in either the vapour or liquid phase. This means that different processes such as methanogenesis, CO_2 production, H_2S production can be measured concomitantly and the ways in which they interact determined quantitatively.
4. Instruments have long-term stability with only infrequent calibration required.
5. Large detection ranges from ppm to several tens of percent are available.
6. Instruments are available that are portable for environmental work.
7. Micro-probes are available which can also be left in an environment for a considerable time allowing long-term measurements at different sites and at different times of the year, thus allowing seasonal effects to be studied.

Previous studies have measured O_2 and CO_2 exchange during photosynthesis by *Scenedesmus* (39), dissolved nitrogen in relation to rates of N_2 fixation as well as methanogenesis during composting (12). More recently mass spectrometry has demonstrated that denitrification, long thought to be a strictly anaerobic process, can occur in the presence of oxygen in both *Paracoccus denitrificans* and *Pseudomonas aeruginosa* (26).

This powerful biophysical technique can therefore be used for measurements of gross fluxes and activities associated with specific groups of microorganisms. Where GEMs are to be used in pure culture, QMS holds great potential for highly sensitive and accurate measurements of activity. In heterogeneous populations it can be used to chart the effects of GEMs on key environmental processes and potentially as an early warning system for any major perturbations. Some potentially useful applications of QMS as an ecological tool have been discussed by Boddy & Lloyd (5).

Another rapid method for detecting active bacteria is by impedance.

Impedance is the resistance to flow of an alternating current as it passes through a conducting material. Thus, insertion of two electrodes into growth medium will give an impedance measurement for that medium; growth of organisms within the culture medium will change the impedance. This change can be electronically monitored to measure growth or the presence of viable organisms as low as 10 cfu/ml. Thus impedance may have a role for rapid detection of GEMs in some highly specialized cases. Impedance methods and equipment have been reviewed by Kell & Davey (22) and Silley (49).

The methods for measuring activity that have been discussed here have great potential in specialized applications of GEMs—particularly in pure culture. They also have considerable scope as policing devices for monitoring any impact of GEMs on the activity of resident indigenous species. However, the inability of detection techniques to accurately enumerate viable cells, including non-culturable species, is a considerable setback for accurately measuring process rates on a per cell basis. That is to say, how active are cells in natural environments? In the final section of this chapter, the technique of flow cytometry will be discussed in some detail. Methods and applications will be described that have potential for overcoming the many disadvantages or deficiencies of the techniques that have been discussed so far.

FLOW CYTOMETRY

Flow cytometry has been used to study biological processes principally during the last 20 years. The bulk of the work has centred on studies of eukaryotes, particularly mammalian cells. It is only in the last 10 years that the instruments have become sensitive enough to be used for studying bacteria. This application has gradually increased so that now flow cytometry holds many exciting possibilities for microbiology and microbial ecology in particular (see Ref. 23). An added facility of some more sophisticated instruments is the ability to select or 'sort out' sub-populations from heterogeneous mixtures.

Flow cytometers combine the advantages of microscopy and biochemical measurements to enable the rapid analysis of thousands of individual cells in seconds. A simple diagram of the basic principles by which these instruments work is shown in Fig. 1.4. Cells are injected into a pressurized stream of liquid, termed the *sheath fluid*, flowing at high speeds (30 m s^{-1}). The result is that cells are diluted out and arranged in a single file within the sheath fluid, which then passes through a focused light beam. The light may be from high-energy arc lamps or lasers. Each cell is presented to the light for a short period of time and two basic properties can be derived for each cell.

First, light will be scattered, the amount of scattering being a reflection of cell size. Scattered light is measured by special detectors. Second, if the cell is stained with a fluorescent dye or molecule then providing the wavelength of the incident light beam can excite the fluorochrome, fluorescence will be emitted which can be quantified by a specific fluorescence detector. Most instruments have the capacity to change the wavelength of the incident light

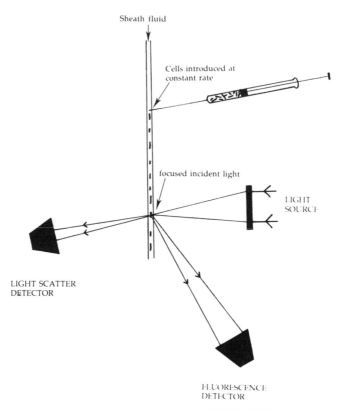

Fig. 1.4 A diagram illustrating some of the principles of flow cytometry.

beam in order to provide a range of excitation wavelengths for a variety of fluorescent molecules.

Figure 1.4 is much simplified; most instruments have two light scatter detectors and two or more fluorescent ones. This makes possible a multiple parameter approach so that for a single cell, low-angle scatter (related to size), high-angle scatter (related to internal structure), and fluorescence emission from two or more *different* fluorescent stains can be measured at the same time. There is a wide range of fluorescent stains available that are specific for different cell structures (DNA, protein) or whose fluorescence is dependent on a specific cellular activity (internal pH, membrane potential).

Light intensity (scattered or fluorescence) measured by the appropriate detectors is converted into electrical signals. These are graded in increasing order of magnitude and assigned to channels, the higher the channel number the greater the light intensity. Thus, running a stream of cells through the sensing region results in each cell emitting a package of light (scatter or fluorescence), the magnitude of which is reflected by the channel number to which it is assigned. Each channel accumulates packages of light from single cells as a number of single events. The final display is in the form of a

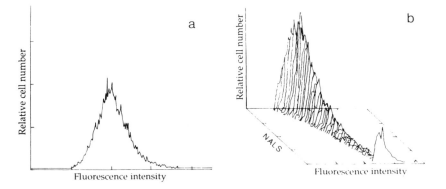

Fig. 1.5 Typical flow cytometric output showing (a) histogram distribution of cell numbers versus fluorescence intensity and (b) two-dimensional display of narrow angle light scatter (NALS)—a measure of cell size, plotted against relative fluorescence intensity within different size classes together with relative cell numbers for each size class distribution. The data were obtained from *Staphylococcus aureus* cells stained with FITC-labelled IgG (from Diaper & Edwards, unpublished).

histogram distribution of numbers of events within a range of channels. Figure 1.5a shows a typical output. Finally, on-line computing allows data manipulation so that:

1. The grading of light intensity into channels can be manipulated so that some channels can be made 'invisible', which means that in mixed populations a sub-population of interest can be studied.
2. Data accumulated by the individual detectors can be stored and then plotted against those obtained by others. For example, light scatter against fluorescence shows two major features: (i) size range of a particular organism as a range of channels, which essentially reflects the cell cycle of that organism; and (ii) the fluorescence intensity associated with each size class of cells, i.e. reflect the amount of the bound component at different stages of the cell cycle (see Fig. 1.5b).

How can this technique be useful for monitoring GEMs in the environment? The next section details some useful applications of flow cytometry and how these are being used, or could be used to monitor a target species within diverse ecosystems.

Applications

Some of the applications of flow cytometry that have been, or are being developed for microorganisms are listed in Table 1.2. DNA-specific dyes such as Hoechst 33342, which binds to adenine-thymidine-rich regions, and mithramycin make it possible to determine chromosome number per cell in bacteria (52). This can yield information regarding physiological status of a cell, e.g. a population of cells having a single chromosome per cell would

Table 1.2 Some cellular structures/activities detectable by flow cytometry.

Target	Example of fluorescent agent	Applications
DNA	Hoechst 33342, mithramycin, ethidium bromide, chromomycin	Chromosome number, studies of the cell cycle
RNA	Pyronin Y, propidium iodide	Physiological status, macromolecular content
Protein	Fluorescein isothiocyanate	
Pigments	F_{420} (deazaflavin), chlorophyll/ phycoerythrin	Methanogens, photosynthetic organisms
Intracellular structures	Light scatter	Detection of PHB— containing *Azotobacter vinelandii*, detection of recombinant bacteria over-synthesizing cloned protein(s)
Enzymes	Fluorescein-, naphthol- and coumaryl-linked substrates	Discrimination of live or damaged cells, following differentiation, quantification of subcellular organelles
Membrane potential	Rhodamine 123	Viability, physiological status
Antigens	FITC-conjugated antibodies	Identification, detection and enumeration of
16S rRNA	FITC-linked oligonucleotides	species within heterogeneous populations

suggest a non-growing state. Major macromolecular pools such as protein (fluorescein isothiocyanate [FITC]) and RNA (Pyronin Y) may be monitored qualitatively by flow cytometry. However, this approach must be ratified by chemical measurements. Recently Kell *et al.* (23) have used cellular auto-fluorescence as an indicator of physiological status, and noted heterogeneity in the amount of fluorescence emitted by cells at different stages of the cell cycle in *Micrococcus luteus*. Because methanogens contain F_{420}, a deazaflav-ine-type compound that is fluorescent, they can readily be detected on the basis of their autofluorescence. Two areas of application of flow cytometry are of particular relevance to tracking and detecting viable GEMs in the environment. These are (i) methods that identify viable cells—irrespective of whether they are culturable—and (ii) the use of fluorescent molecular probes that allow highly specific identification of a target GEM. Other determinants listed by Kell (23) for flow cytometric analysis include photosyn-thetic pigments (chlorophylls, phycoerythrin), enzyme activities (using substrates linked to fluorescent molecules such as coumaryl or umbelliferyl groups), internal pH, calcium levels and inclusion bodies such as PHB.

Enumeration of viable cells

The problem of cells adopting viable, but non-culturable states in natural environments has been alluded to throughout this chapter. This phenomenon makes it extremely difficult to predict with any degree of certainty the *absence* of any target organism, such as a GEM, from any natural environment. Attempts to identify such forms have been devised, such as microscopic analysis of acridine orange-stained cells. Live cells stain green, and debris (and presumably dead cells) appear orange to red. However, all such attempts have a number of drawbacks that make them unreliable (35,38). The problem is illustrated in Fig. 1.6. *Aeromonas salmonicida* released into filtered seawater was enumerated by flow cytometry (total counts) and plating onto nutrient agar (cfus = colony forming units). As can be seen, total counts remained constant whereas cfus fell rapidly after 8 days. Two explanations may be advanced that demonstrate the dilemma facing researchers attempting to predict the fate of GEMs in the environment.

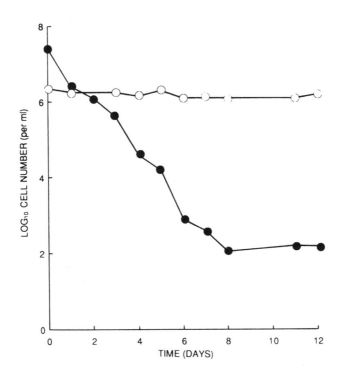

Fig. 1.6 Survival of *Aeromonas salmonicida* in filtered seawater measured by cfus after plating onto nutrient agar (●) or enumerated as a total count by flow cytometry (○) (from Diaper & Edwards, unpublished).

1. Decreased cfus are a true indication of the loss of viability; the unchanging flow cytometer counts reflect a mixture of dead plus live cells.

or

2. The disparity between the numbers obtained by each method is due to a sub-population of viable but non-culturable cells together with another sub-population of dead cells—the exact proportion of these sub-populations being indeterminate.

The second explanation is now gaining much support, particularly from experiments that investigate the physiology and biochemistry of cells in nutrient-limited conditions, some of which have been described at the beginning of this chapter.

A number of fluorescent dyes have been reported that are taken up into eukaryotic cells in response to membrane potential—a property of living cells only. These include rhodamine 123 (10) and 3, 3-dihexyloxacarbocyanine iodide ($DiOC_6$) (43). A study whereby these dyes, as well as fluorescein diacetate, which is retained by cells having intact membranes, were tested against a range of bacterial species for their ability to stain viable cells showed that no single dye had universal application, but that rhodamine 123 was most generally useful (13). An example of the power of flow cytometric analysis to measure viability using rhodamine-stained *Escherichia coli* is shown in Fig. 1.7. *E. coli* cells were stained with rhodamine 123 and could be identified as a fluorescent population displayed as a histogram distribution. Treatment with gramicidin, a membrane-disrupting agent which abolishes membrane potential, almost totally eliminated cell fluorescence (Fig. 1.7a). Mixtures of live:dead cells of *E. coli* were also set up and stained with rhodamine. Live cells had much higher fluorescence than dead ones and could be clearly discriminated (Fig. 1.7b). Finally, an exponential culture of *S. aureus* was diluted to give different cell densities and viability measured using rhodamine staining and flow cytometry and compared to cfus determined from plate counts. Fig. 1.7c shows that numbers were essentially the same by either method.

Other species analysed in this way gave similar results. Therefore flow cytometry holds great promise for identifying and enumerating the numbers of cells that are viable. However, for a GEM, another tool is required before viability may be measured in heterogeneous populations. A means of discriminating a target species has to be available. Again, flow cytometric methods are available which can achieve this.

Enumeration of specifically-labelled cells

Molecular genetic methods are now increasingly directed towards studies of microbial ecology. Applications include understanding population genetics as well as developing molecular probes that allow specific identification down to the level of a selected strain of a species. Of the fluorescent molecular probes available, two have emerged as exquisite molecular tools that may be

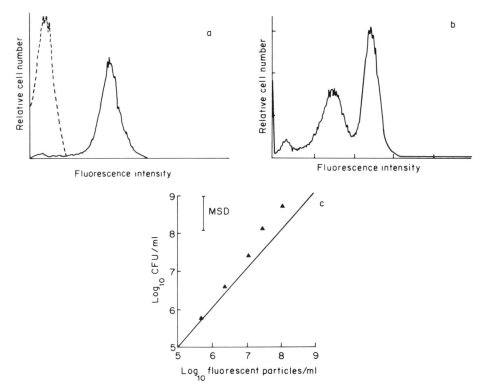

Fig. 1.7 Flow cytometric discrimination of viable cells of *Escherichia coli* using rhodamine 123 (rh123). (a) Viable cells are fluorescent due to uptake of rh123 in response to membrane potential (———); agents that abolish membrane potential (e.g. gramicidin) reduce the fluorescence (-----). (b) Fluorescence histograms of a mixed culture consisting of viable and formaldehyde-fixed (dead) cells of *E. coli* stained with rh123. (c) Relationship between viable plate counts and counts of rh123-stained *S. aureus* determined by flow cytometry. The ideal relationship is represented by the line, (▲) represent actual values. MSD = minimum significant difference (from Diaper & Edwards, unpublished).

applied to identify target organisms. The first involves the use of fluorescent antibodies, which are extremely sensitive and are reported to have lower limits of detection around 2×10^{-1} cells per ml water or per g soil (37). We have tested this approach by allying antibody binding to flow cytometric detection. *Staphylococcus aureus* produces protein A, a cell wall immuno-globulin-binding protein (18). This has been used as a model for fluorescently (FITC) labelled IgG detection and enumeration of *S. aureus* by flow cytometry. *S. aureus* could be detected after binding IgG-FITC followed by flow cytometry. Lower limits of detection were found to be around 10^{1}–10^{2} cells ml^{-1}. *S. aureus* was also released into eutrophic lakewater that contained a heterogeneous population of bacteria that totalled around 10^{7} ml^{-1}. Figure 1.8 shows that the IgG-FITC probe detected *S. aureus* with a high degree of

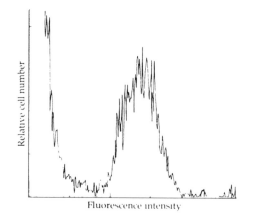

Fig. 1.8 Fluorescence histograms of *Staphylococcus aureus* stained with FITC-IgG after inoculation into lakewater that contained a heterogeneous bacterial population. (from Diaper & Edwards, unpublished).

specificity. The numbers calculated from the fluorescent distribution were essentially the same as for cfus, but the lower limit of sensitivity in this non-sterile system was raised to around 10^3 cells ml^{-1}. However, this could easily be improved by concentration of water prior to testing such that 1 cell ml^{-1} could easily be detected after 10^3–10^4-fold concentration of water by centrifugation or filtration.

However, a note of caution has to be introduced here. As noted at the beginning of this chapter, it is not known what effects prolonged residence in natural environments have on antigenicity.

The second fluorescent molecular probe that shows much promise in specific identification involves fluorescent oligonucleotide probes against intracellular 16S rRNA (1). 16S rRNA allows not only classification of microorganisms into the domains Archaea, Bacteria and Eucarya (58) but contains sufficient variable regions to allow discrimination down to the species level and possibly strains of a single species. Specific identification of individual cells by flow cytometry using this approach has been described by Amann *et al.* (1). A problem that needs to be resolved is the low fluorescent intensity of signals obtained from cells isolated from natural environments. This no doubt reflects low rRNA copy number due to non-growing starvation conditions, which have been discussed at the beginning of the chapter (see also Table 1.1); this drawback is presently exacerbated by the fact that only one dye molecule can be bound per probe. Recently Zarda *et al.* (59) have attempted to improve the amount of fluorescence by labelling oligonucleotides with digoxigenin (DIG), which after probe hybridization can be detected using fluorescently labelled antibodies. Since the binding proteins can be labelled with several fluorescently labelled molecules, an increase in sensitivity should be detected. However, early work is equivocal; some bacteria do indeed show an enhanced fluorescent signal whilst others do not (59).

Where does this leave us? Flow cytometry holds a great deal of promise in analysing target organisms within heterogeneous populations with respect to viability and detection by fluorescent molecules. A promising way forward may be a combination of viability staining and fluorescent antibody or 16S rRNA oligonucleotide probes. In this way, the target GEM may be identified as a discrete sub-population using the probe, and staining with a dye taken up only in response to membrane potential (viable cells) at the same time as probe hybridization can then be used to enumerate the numbers of this sub-population which are dead or viable. This flow cytometric approach can be coupled to traditional plating methods for enumeration of cfus. If viable flow cytometer numbers are significantly different to those obtained by plating, then it is evidence for a degree of non-culturability.

The model systems described here point the way forward for overcoming many of the problems we posed at the beginning of this chapter for the unambiguous identification and enumeration of GEMs from natural environments. However, this approach will depend on the success of extraction of viable organisms from diverse ecosystems, and this is dealt with elsewhere in the book. On a wider front, the coupling of molecular genetic techniques to sophisticated biophysical instrumentation holds much promise in the area of monitoring GEMs and other bacteria in natural environments as well as in more traditional areas of microbial physiology, pathogenicity and biotechnology. No doubt these emerging technologies will force us to reassess our understanding of microorganisms, which has largely come from the more conventional and artificial conditions of the laboratory.

ACKNOWLEDGEMENT

Some of the work described in this chapter was funded through grants from the NERC.

REFERENCES

1. Amann, R.I., Krumholz, L. & Stahl, D.A. (1990) *Journal of Bacteriology* **172**, 762–770.
2. Amann, R.I., Binder, B., Chisholm, S.W., Olsen, R., Devereux, R. & Stahl, D.A. (1990) *Applied and Environmental Microbiology* **56**, 1919–1925.
3. Amann, R.I., Stromley, J., Devereux, R., Key, R. & Stahl, D.A. (1992) *Applied and Environmental Microbiology* **58**, 614–623.
4. Binnerup, S.J., Jensen, K., Reusbech, N.P., Jensen, M.H. & Sorenson, J. (1992) *Applied and Environmental Microbiology* **58**, 303–313.
5. Boddy, L. & Lloyd, D. (1989) In *Field Methods in Terrestrial Ecology Nutrient Cycling*, Eds Harrison, A.F. & Ineson, I.P., Springer-Verlag, Berlin, p. 139.
6. Boddy, L., Morris, C.W. & Wimpenny, J.W.T. (1990) *Binary* **2**, 179–185.
7. van Bogelen, R.A. & Neidhardt, F.C. (1990) *FEMS Microbiology Ecology* **74**, 121–128.

8. Chesbro, W., Arbige, M. & Eifert, R. (1990) *FEMS Microbiology Ecology* **74**, 103–120.
9. Colwell, R.R., Brayton, P.R., Grimes, D.J., Roszak, D.R., Huq, S.A. & Palmer, L.M. (1985) *Biotechnology* **3**, 817–820.
10. Darzynkiewicz, Z., Traganos, F., Stainio-coico, L., Kapusainski, J. & Melamed, M.R. (1982) *Cancer Research* **42**, 799–806.
11. De Long, E.F., Wickham, G.S. & Pace, N.R. (1989) *Science* **243**, 1360–1363.
12. Derikx, P.J.L., de Jung, G.A.H., Opden Camp, H.J.M., van der Drift, C., van Griensven, L.J.L.D. & Vogels, G.D. (1989) *FEMS Microbiology Ecology* **62**, 251–258.
13. Diaper, J.P. & Edwards, C. (1992) *Applied Microbiology and Biotechnology* (in press).
14. Forbes, K.J., Bruce, K.D., Jordens, J.Z., Ball, A. & Pennington, T.H. (1991) *Journal of General Microbiology* **137**, 2051–2058.
15. Ford, S. & Olson, B.H. (1988) *Advances in Microbial Ecology* **10**, 45–79.
16. Gottschal, J.C. (1990) *FEMS Microbiology Ecology* **74**, 93–102.
17. Goullet, P. & Picard, B. (1988) *Journal of General Microbiology* **134**, 317–325.
18. Harlow, E. & Lane, D. (1988) *Antibodies: a Laboratory Manual.* Cold Spring Harbor Laboratory, USA.
19. Helm, D., Labischinski, H. & Naumann, D. (1991) *Journal of Microbiological Methods* **14**, 127–142.
20. Helm, D., Labischinski, H., Schallehn, G. & Naumann, D. (1991) *Journal of General Microbiology* **137**, 69–79.
21. Jain, R.K. & Sayler, G. (1987) *Microbiological Sciences* **4**, 59–63.
22. Kell, D.B. & Davey, C.L. (1990) In *Biosensors: a Practical Approach*, Ed. Cass, A.E.G., Oxford University Press, Oxford, pp. 125–154.
23. Kell, D.B., Ryder, M.M., Kaprelyants, A.S. & Westerhoff, H.V. (1991) *Antonie van Leewenhoek* **60**, 145–158.
24. Kramer, J.G. & Singleton, F.L. (1992) *Applied and Environmental Microbiology* **58**, 201–207.
25. Lloyd, D. & Scott, R.I. (1983) *Journal of Microbiological Methods* **1**, 313–328.
26. Lloyd, D., Boddy, L. & Davies, K.J.P. (1987) *FEMS Microbiology Ecology* **45**, 185–190.
27. van Loosdrecht, M.C.M., Lyklema, J., Norde, W. & Zehnder, A.J.B. (1990) *Microbiological Reviews* **54**, 75–87.
28. McKay, A.M. (1992) *Letters in Applied Microbiology* **14**, 129–135.
29. Magee, J.T., Hindmarch, J.M., Duerden, B.I. & Mackenzie, D.W.R. (1988) *Journal of General Microbiology* **134**, 2841–2847.
30. Matin, A. (1990) *FEMS Microbiology Ecology* **74**, 185–196.
31. Mekalonos, J.L. (1992) *Journal of Bacteriology* **174**, 1–7.
32. Nichols, P.D., Henson, J.M., Guckert, J.B., Nivens, D.E. & White, D.C. (1985) *Journal of Microbiological Methods* **4**, 79–94.
33. Norris, K.P. (1959) *Journal of Hygiene* **57**, 326–345.
34. Nystrom, T., Abertson, N.H., Flardh, K. & Kjelleberg, S. (1990) *FEMS Microbiology Ecology* **74**, 129–140.
35. Page, S. & Burns, R.G. (1991) *Soil Biology and Biochemistry* **23**, 1025–1028.
36. Picard-Pasquier, N., Picard, B., Heeralal, S., Krishnamoorthy, R. & Goullet, P. (1990) *Journal of General Microbiology* **136**, 1655–1666.
37. Pickup, R.W. (1991) *Journal of General Microbiology* **137**, 1009–1019.
38. Postma, J. & Altenuller, H.J. (1990) *Soil Biology and Biochemistry* **22**, 89–96.

39. Radmer, R. & Ollinger, O. (1980) *Plant Physiology* **65**, 723–729.
40. Rajendran, N., Matsuda, O., Imamura, N. & Urushigawa, Y. (1992) *Applied and Environmental Microbiology* **58**, 562–571.
41. Roszak, D.B. & Colwell, R.R. (1987) *Microbiological Reviews* **51**, 365–379.
42. Scopes, A.W. (1962) *Journal of General Microbiology* **28**, 69–79.
43. Shapiro, H.M. (1988) In *Practical Flow Cytometry*, Alan R. Liss, New York.
44. Shaw, J.J. & Kado, C.I. (1986) *Bio/Technology* **4**, 560–564.
45. Shaw, J.J., Davie, F., Geiger, D. & Kloepper, J.W. (1992) *Applied and Environmental Microbiology* **58**, 267–273.
46. Shotton, D.M. (1990) *Methods in Cell Biology* **33**, 2–35.
47. Sieracki, M.E., Johnson, P.W. & Sieburth, J.McN. (1985) *Applied and Environmental Microbiology* **49**, 799–810.
48. Sieracki, M.E., Reichenbach, S.E. & Webb, K.L. (1989) *Applied and Environmental Microbiology* **55**, 2762–2772.
49. Silley, P. (1991) *Society of General Microbiology Quarterly* **18**, 48–52.
50. Spector, M.P. (1990) *FEMS Microbiology Ecology* **74**, 175–184.
51. Stead, D.E., Sellwood, J.E., Wilson, J. & Viney, I. (1992) *Journal of Applied Bacteriology* **72**, 315–321.
52. Steen, H.B., Skarstad, K. & Boye, E. (1990) *Methods in Cell Biology* **33**, 519–526.
53. Tempest, D.W. & Neijssel, O.M. (1984) *Annual Reviews of Microbiology* **38**, 451–486.
54. Viles, C.L. & Sieracki, M.E. (1992) *Applied and Environmental Microbiology* **58**, 584–592.
55. Watson, S.W., Novitsky, T.J., Quinby, H.L. & Valois, F.W. (1977) *Applied and Environmental Microbiology* **33**, 940–946.
56. White, D.C. (1983) In *Microbes in their Natural Environments'*, Eds Slater, J.H., Whittenbury, R. & Wimpenny, J.W.T., Society of General Microbiology Symposium No. 34. Cambridge University Press, Cambridge, pp. 37–66.
57. White, N. (1991) *Society of General Microbiology Quarterly* **18**, 70–74.
58. Winker, S. & Woese, C.R. (1991) *Systematic and Applied Microbiology* **14**, 305–310.
59. Zarda, B., Amann, R., Wallner, G. & Schleifer, K.H. (1991) *Journal of General Microbiology* **137**, 2823–2830.

Chapter 2

Genotypic and Phenotypic Methods for the Detection of Specific Released Microorganisms

J.R. Saunders[1] and V.A. Saunders[2]
[1]Department of Genetics & Microbiology, University of Liverpool
[2]School of Biomolecular Sciences, Liverpool John Moores University

INTRODUCTION

The proposed release of genetically engineered microorganisms (GEMs) into the open environment has prompted a need to monitor the fate of specific microorganisms and the genes they carry. GEMs are usually the products of a range of both *in vivo* and *in vitro* gene manipulations. It is therefore fitting that molecular genetic techniques are ideally suited to the problem of tracking organisms following release and of determining the potential fate of any recombinant DNA molecules that they carry. Methods are needed that permit the detection of specific released organisms and their recombinant genes when present at very low concentrations in natural environments. Since the recombinant genes may be transmitted to indigenous organisms in

Monitoring Genetically Manipulated Microorganisms in the Environment. Edited by C. Edwards
Published 1993 John Wiley & Sons Ltd. © 1993 J.R. Saunders and V.A. Saunders

the environment it is essential that the released host can be distinguished from the natural residents. Furthermore, since microorganisms may become dormant whilst in the environment, detection methods should ideally be able to provide a measure of the biological activity of specific organisms.

This chapter outlines technologies that are currently in use or under development for the detection, identification, enumeration and activity measurement of specific genetically engineered bacteria. Methods for the detection of individual genes, gene products and release host organisms themselves are described. Most of the examples given refer to bacterial systems, but the techniques involved are generally applicable to genetically engineered fungi and other microorganisms.

MARKER SYSTEMS

The detection and identification of specific microorganisms in natural habitats is greatly facilitated by the carriage of readily identifiable and, ideally, unique combinations of genes (40). In many cases the traits engineered into a GEM to provide the functional qualities for which it is to be released will be unsuitable for purposes of detection and monitoring. For example, an engineered organism destined for release may be one whose sole difference from the wild-type is the *absence* of a particular genetic function, as in the construction of *ice*-minus strains of *Pseudomonas syringae* (85). The detection of strains with such negative characteristics may be precluded by a background of wild-type and unrelated bacteria. The addition of indicator genes with readily identifiable phenotypes greatly assists in the tracking of functional GEMs and is a crucial ingredient of model GEMs used to predict behaviour in natural environments.

Marker genes and sequences

A variety of indicator genes, including those for antibiotic or heavy metal resistances from various bacteria (e.g. 82, 170, 171), β-galactosidase (*lacZ*) from *Escherichia coli* (36), β-glucuronidase (*gusA*) from *E. coli* (66), thymidylate synthase (*thyA*) from *Lactococcus lactis* (122), catechol 2,3 dioxygenase (*xylE*) from *Pseudomonas putida* (171) and luciferase (*lux*) from *Vibrio* species (121, 134), have been used for *in vivo* and *in situ* studies of bacteria. Where a background of naturally occurring genes for any particular marker is found in the ecosystem an appropriate combination of markers may enhance selectivity.

Resistance genes have the advantage that they permit direct selection from environmental samples provided that the bacteria carrying them can be cultured. There may, however, be problems with the use of certain resistance markers due to their frequent distribution in some populations (70). Furthermore, the conditions of nutrient stress encountered in natural habitats may produce phenotypic resistance in the absence of specific resistance genes.

There are also ethical considerations about releasing yet more resistant bacteria into the environment in view of possible problems associated with increasing the already high load of resistance genes in bacterial populations. However, it could be argued that the distribution of drug resistance in clinically important bacteria is already so high, due to the use and abuse of antibiotics, that any increase in incidence resulting from the deliberate release of GEMs is unlikely to have any significant adverse impact.

Most of the alternative indicator genes that have been developed lack the selective advantages of resistance genes, but their presence may be detected—colorimetrically by using efficient enzyme assays, fluorimetrically by using suitable fluorochromes or by measuring bioluminescence—and may permit direct detection *in situ*.

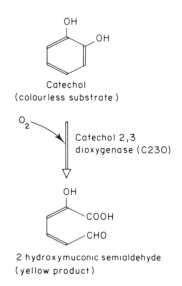

Catechol
(colourless substrate)

O$_2$

Catechol 2,3
dioxygenase (C230)

OH

COOH

CHO

2 hydroxymuconic semialdehyde
(yellow product)

Fig. 2.1 Catechol 2,3 dioxygenase reaction: degradation of catechol by catechol 2,3, dioxygenase results in a yellow coloured product.

The *xylE* gene of *Pseudomonas putida* encodes the production of catechol 2,3 dioxygenase (C230), which converts catechol to 2-hydroxymuconic semi-aldehyde (Fig. 2.1), a yellow coloured product that can be assayed readily. Colonies expressing the *xylE* product and the enzyme itself can be detected since they develop a yellow coloration after being sprayed with a solution of catechol (1% w/v). C230 has the useful and relatively uncommon property of being stable in acetone. This property greatly assists the purification and assay of the enzyme from environmental samples (171). The enzyme is also denatured when exposed to air in the absence of acetone. Thus, C230 released from lysed cells will rapidly become inactivated in natural aerobic environments. This means that C230 activity can be used as a measure of the fraction of host bacteria that are still viable (or at least unlysed). Even so, inactivated enzyme can be detected immunologically (102). The *xylE* gene in combination with various promoters has been incorporated into both broad host range non-conjugative and conjugative plasmids as well as the chromosome of model release host strains of *P. putida* and has been used to monitor the fate of recombinant pseudomonad populations in lakewater (171–173) and soil (89). *xylE* marker plasmids have also been developed for use in Gram–positive bacteria such as *Streptomyces* spp. (174).

The *E. coli lacZY* genes, encoding β-galactosidase and lactose permease, respectively, provide a marker system for use in organisms lacking the *lac* operon and incapable of using lactose as sole carbon source. Presence of the marker is determined by cleavage of Xgal (5-chloro-4-bromo-3-indolyl β-D-galactopyranoside) by β-galactosidase and production of a blue colour. This marker system permitted the detection of fluorescent pseudomonads at a sensitivity of <10 cfu per g soil (36). However, there is a requirement for culturing the host on solid medium with this system.

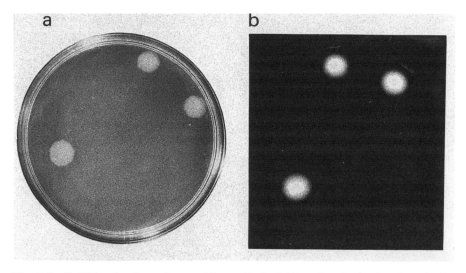

Fig. 2.2 Self-illumination of recombinant *Escherichia coli* colonies bearing the lux indicator system. (a) *E. coli* HB101 carrying a *lux* indicator vector was grown on a nutrient agar plate overnight at 37°C. (b) The plate was exposed in total darkness to Kodak T-MAX P 3200 PROF. film (uprated to 12 500 ASA) for 45 min at f5.6. (Courtesy of Dr R.W. Pickup and Mr B. Simon, Institute of Freshwater Ecology, Windermere.)

Luciferase genes (*lux*), notably the *luxB* gene from *Vibrio fischeri*, provide versatile reporter genes for monitoring the presence and activity of bacteria in a variety of environments (121,135,148). Bacteria carrying such genes can be identified by their bioluminescence provided they can generate sufficient energy to drive photon emission (Fig. 2.2). The majority of energy is provided by the oxidation of a long-chain aliphatic aldehyde which can be generated intracellularly in the presence of the *luxCDE* genes or provided externally as dodecanal. However, expression of the bioluminescence pheno-type also depends on the ability of cells to produce $FMNH_2$, which is supplied in aerobic bacteria via an NADH/FMN couple and the electron transport chain. Global changes in the intracellular physiology of bacteria caused, for example, by environmental stresses, may affect the output of light making the system unreliable for quantitative monitoring of bacteria in the environ-ment. The presence of indigenous luminescent bacteria in aquatic environ-ments and of silent *lux* gene sequences may also jeopardize the utility of *lux* as a marker in aquatic environments.

Further approaches to the detection of such indicator genes or gene products include nucleic acid hybridization and immunological methods (see later sections). Some marker systems utilize unique synthetic oligonucleotide signatures or heterologous DNA sequences detectable by using specific hybridization probes, sometimes in conjunction with *in vitro* amplification of target DNA (see later section, p. 39). Bishop and co-workers (18) synthe-sized an 80 bp DNA sequence that contained translation stop codons in all three reading frames on both strands, but no ATG start codon and no other signals likely to affect replication or gene expression of the host genome. This served as a marker incorporated into a dispensible intergenic, non-regulatory region of the genome of an occluded baculovirus of *Autographa californica* (AcNPV) for the first field release study of a genetically engin-eered organism undertaken in Britain. For bacteria, marker DNA sequences of eukaryotic origin, such as a genomic sequence from napier grass (27; see also later section, p. 39), that are not homologous to the DNA from the ecosystems under test, can be employed to discriminate between GEMs and the indigenous microflora.

Limited survival is essential for most released organisms to carry out their alloted task. Complete physical containment of organisms released for agricultural or environmental detoxification purposes cannot be achieved. Therefore, it has been proposed that release strains should contain suicide genes to ensure that the released population dies out (14), for example in response to external stimuli such as ultraviolet light or temperature. One such suicide system involves the *hok* gene derived from plasmid R1 of *E. coli* and which encodes a 52 amino acid peptide that is lethal when expressed in a variety of bacterial species (101). By coupling the *hok* gene to the randomly invertible *fimA* promoter of *E. coli* (Fig. 2.3) it has proved possible to produce a lethal gene cassette that when carried by a released GEM would lead to elimination of the organism over a limited period of time (101). However, due to the likelihood of deletion events or gene transfer a single system of suicide genes will probably not be sufficient and some degree of redundancy, in the form of two or more independent lethal genes, will need to be carried by strains released into the open environment. In addition to providing a suicide function, such genes also provide targets for identifying the GEM in the environment.

The inherent attributes of a particular organism may also be exploited as

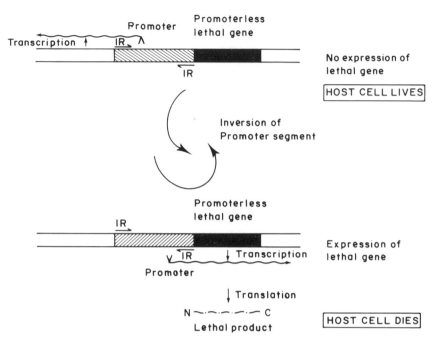

Fig. 2.3 Random expression of a lethal gene by inversion of a control region carrying a promoter. A promoter lies in an invertible control region upstream of the promoterless lethal gene. Inversion of this region involves recombination between the inverted repeats. In one orientation the promoter directs transcription away from the lethal gene, which is thus not expressed. In the inverted orientation the promoter is directed towards the lethal gene which is thus transcribed and subsequently translated. Inversion can occur randomly in each generation. IR, inverted repeat; N, amino, and C, carboxy end of polypeptide.

identity tags. For example, a genetic fingerprint of an organism may be used. Specific DNA fragments, generated when genomic DNA is digested with restriction enzymes, may be highlighted by hybridizing with complementary oligonucleotide probes to produce a profile of hybridizing fragments that is characteristic of a particular organism (see, for example, Ref. 51). The availability of appropriate hybridizing probes complementary to species-specific sequences of ribosomal RNA (rRNA) permits the detection and characterization of target organisms in environmental samples (see later section, p. 42). Biochemical signatures, based for example on metabolic potential (19), or multilocus enzyme electrophoresis (133) offer a further means of tracking and identifying released GEMs and bacteria to which their genes have been transferred. However, such approaches require both the culture of environmental isolates and an extensive databank of specific signatures for comparison and identification.

The problem of expression

Indicator genes may be engineered to achieve optimum, or at least acceptable, levels of expression within a release host. This may be achieved by using either known strong promoters of heterologous origin, such as the *lac* promoter (36) or the leftward or rightward promoters of coliphage lambda (171), or strong promoters indigenous to the release host. For phenotypic detection of gene products it is also necessary that the appropriate translational signals are present to ensure correct gene expression. Strong expression may, however, be deleterious to the release host (or to any subsequent organism to which an indicator gene might be transferred) by imposing an excessive physiological burden on the host and potentially reducing fitness. This problem can be at least partially overcome by the use of controllable promoters to direct expression of the indicator gene. If the conditions required to produce maximal expression are unlikely to be achieved under normal environmental conditions for the GEM, then the physiological burden can be reduced. For example, lambda P_L and P_R can be regulated by the cI_{857} gene which encodes a temperature-sensitive repressor for these promoters. At temperatures below 30°C this repressor functions normally to shut off expression from these promoters very tightly. However, at 37°C and above the repressor becomes thermally inactivated and expression ensues. Such high temperatures are unlikely to be achieved in most habitats into which GEMs are released. Therefore expression of the indicator gene may be restricted to the period of post-sampling, which reduces any burden on the release host in the open environment. Furthermore, the specificity of marker genes is enhanced (particularly if there is a background of naturally occurring organisms with the same or similar genes) by virtue of the demonstrable ability to direct marker expression in the GEM, at will, by manipulation of incubation conditions. Similar approaches can be used with other promoter systems such as P_{trp}, that respond to particular metabolites or such factors as UV irradiation or temperature.

Positional effects

In view of the potential problem of gene transfer from released microorganisms (see later section, p. 48) it seems likely that many GEMs will be engineered to contain marker and/or operationally significant genes in the chromosome rather than on recombinant plasmids. Whilst location on the chromosome may reduce the possibility of transfer it cannot, however, be totally eliminated. A released strain could, for example, subsequently become infected with a conjugative plasmid that has chromosome mobilizing ability (cma). Strategies can be envisaged whereby release strains, e.g. soil inoculants, would be generally marked at one or more sites on the host chromosome. A chromosomal position would circumvent the stability problems associated with plasmid segregational loss. However, chromosomal insertions mean that the copy number advantages of a plasmid location, in relation to gene dosage effects on detection and/or function, might be foregone. Some compensation for this loss may be achieved by using efficient, controllable promoters inserted upstream of the chromosomal indicator genes.

Chromosomal insertion of indicator genes presents several problems for GEMs, not least of which is the potential inactivation of functional genes and consequent reduced fitness of the release host. Furthermore, the position of insertion may interfere with the differential transcription of genes located on different supercoiled domains of the folded chromosome. The position of insertion might also determine the relative accessibility of cma activity of conjugative plasmids and thus affect the likelihood of dissemination of engineered genes.

Marker systems are obviously employed to help monitor the fate of individual GEMs and the genes contained therein. However, introduced indicator genes or sequences may prove to be unstable and may reduce the fitness of the GEM. It could be argued that the introduction of such genes into release hosts is unnecessary, since the only genes that would be significant to a 'working' GEM would be those that affect the functional properties of the released organism. However, such properties may not provide the best markers, particularly if the engineering involves the loss or subtle alteration of the activity of the gene(s).

DETECTION TECHNIQUES

Methods involved in the detection and identification of GEMs are in essence no different from those involved in detecting other microorganisms, in particular pathogenic bacteria such as *Salmonella* species in foodstuffs or other environments. Indeed there is likely to be a two-way exchange of technologies in these areas.

The problem of non-culturable microorganisms

It is essential to be able to identify the components of microbial communities and determine their activities in order to establish whether any perturbations could arise as a consequence of deliberate introductions of manipulated organisms (116). Unfortunately, little is known about the species composition and genetic diversity of natural bacteria communities, largely due to the fact that, at best, only about 1% of the bacteria present can be cultured using available techniques (68, 79). This arises either because current culture techniques are inadequate to recover all species present, particularly those inhabiting low nutrient environments (65), or because some bacteria, notably pathogens such as *Vibrio cholerae* (30), *Legionella pneumophila* (63) and *Salmonella* species (71), can enter a non-culturable but still viable state. It is evident, therefore, that the majority of genetic diversity lies amongst the unculturable fraction of most microbial communities in both aquatic and terrestrial environments (49, 157, 165). This fraction may act as a 'sink' into which particular engineered genes might escape and be sequestered from detection by conventional culture-based techniques. The non-culturable fraction could also act as a reservoir for extrachromosomally-borne genes that might be transferred to released organisms with as yet unknown consequences. In this respect it has been shown that non-culturable *E. coli* cells can maintain the multicopy plasmid pBR322 over long periods (26). Despite these problems culture techniques continue to be widely used in environmental microbiology. Such techniques simply measure the number of colony-forming units (cfu) present when a cell, collection of cells, spore or a fragment of mycelium is allowed to form a colony on the surface of an appropriate solid medium. It is possible to improve these techniques by observing cell growth and/or division microscopically in microcolonies (see, for example, 43,154). For many species these culture methods continue to be easy and cheap to perform and are both sensitive and amenable to statistical analysis (29).

Bacterial growth rates and productivity in natural environments have been estimated by measuring the incorporation of [³H]thymidine into DNA (44). Productivity can be calculated for both free-living and attached bacteria using this technique (64). By determining the uptake of the radiolabelled DNA precursor using microautoradiography as many as 90% of total bacteria have been shown to be metabolically active in some environments (44). This further illustrates the problem of only analysing the culturable fraction of the bacterial population within a community. There are, however, obvious problems of using such techniques for monitoring the fate of GEMs, notably lack of specificity, variations in the metabolism of nucleic acid precursors and differential lysis and recovery of DNA.

Inability to obtain pure cultures of the vast majority of bacteria has proven a major obstacle to our understanding of even simple microbial communities. Direct observation of bacterial cells by microscopy avoids the requirement for culture. The use of acridine orange as a fluorescent stain for nucleic acids in bacteria has provided a means of direct detection and of obtaining total counts in environmental samples (42, 60, 69). There are, however, problems associated with the use of acridine orange. These include both non-specific staining and a high background. Such problems have been overcome by utilizing alternative fluorochromes such as bisbenzimide (17), 4′,6-diamidino-2-phenylidone (DAPI) (117) or rhodamine 123 (95). A procedure for estimating viable counts is to observe the elongation of cells when incubated in low nutrient conditions in the presence of the DNA gyrase inhibitor, nalidixic acid (73). In this technique any cell which demonstrates elongation constitutes a viable unit, since inhibited cells continue to grow but fail to divide. The problem with this technique is that the elongation response is not seen with cocci and the presence of cells of varying size and shape in natural mixed bacterial communities confuses measurement (1). Obviously only a limited number of phenotypic characters may be observed microscopically in this way. For a variety of reasons, therefore, there has been a great impetus to develop techniques for the detection and identification of specific bacteria by means that do not rely on prior culture or on the limited metabolic activity necessary to produce, for example, elongation in the presence of nalidixic acid. The use of such techniques will not only avoid problems of non-culturability but will also be likely to lessen the time delay in obtaining results due to the often slow growth of some bacteria in laboratory culture. Newly developed nucleic acid probe assays and immunoassays not only offer increased speed but also increased specificity when compared with conventional methodologies based on culture.

Nucleic acid-based detection

DNA isolated directly from bacteria in environmental samples provides genetic information about both the culturable and non-culturable fraction. The structurally related fluorochromes DAPI and Hoechst 33258, can be used to determine the amounts of DNA in environmental samples (24,112). Although total DNA measurements will include extracellular DNA released from dead cells, estimates of cell numbers based on DNA contents seem to be in general agreement with those obtained by acridine orange direct counting (112).

Diversity in bacterial communities can be assessed on the basis of the heterogeneity of the isolated DNA. Assessment has normally depended on methods that reveal phenotypic features of the community, but these present a number of problems. One of these is a requirement for the isolation and cultivation of the bacterial strain involved, which thereby excludes the non-culturable fraction of the community. Another is that by analysing phenotypic features only a restricted part of the genome is explored.

To provide a more complete picture of the genome and to gain genetic information about the non-culturable bacteria, DNA heterogeneity of the community has been estimated by determining the reassociation kinetics of the total DNA following its denaturation (157, 158). Spectrophotometric measurement of reassociation kinetics of homologous single DNA strands gives a measure of the genome size and of genetic complexity of bacteria. The complexity of a genome is defined as the number of nucleotides per haploid genome, assuming that it does not contain repetitive DNA (23). The fraction of renatured DNA is expressed as a function of the product (C_0t) of the DNA nucleotide concentration (C_0), in moles/l and the reaction time (t), in seconds. Under appropriate conditions the $C_0t_{1/2}$ (C_0t for the half-completed reaction) is proportional to the genome size or complexity of the DNA (Fig. 2.4) that represents the genome of an organism. The genetic diversity of a microbial community can be expressed in a similar way by using the $C_0t_{1/2}$ value as an index of diversity (157, 158). The $C_0t_{1/2}$ value obtained for soil by Torsvik and co-workers (157) was about 4600, which corresponds to approximately 4000 different bacterial genomes. Reassociation of homologous single-stranded DNA follows second-order kinetics. However, the reassociation curves for the denatured soil bacterial DNA did not exhibit

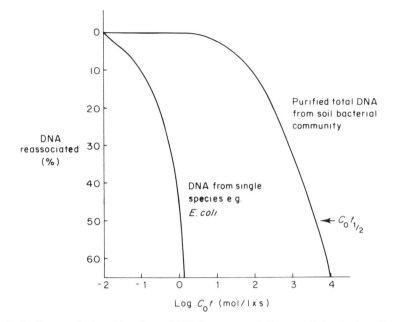

Fig. 2.4 Reassociation kinetics of DNA extracted from soil bacteria. $C_o t$ is the product of the DNA nucleotide concentration (C_o) in mol/l and the reaction time (t) in seconds. $C_o t_{1/2}$ is $C_o t$ for the half-completed reaction.

ideal second-order kinetics indicating that there are groups of both common and rare genomes in the soil community. This complicates precise measurement of sequence heterogeneity. Nevertheless, measurements of genetic diversity by $C_0 t_{1/2}$ analysis are in good agreement with estimates of diversity based on phenotypic characteristics (158).

Extraction of nucleic acids from environmental samples

Hybridization and other methods for the molecular analysis of nucleic acids generally depend on the ability to obtain DNA or RNA from environmental samples. Some nucleic acids are found extracellularly in natural aquatic and soil environments (111). The dissolved DNA content in lake and river water ranges from about 1 to 300 ng ml^{-1} (in contrast, RNA is present at <0.1 ng ml^{-1}) with turnover times of the order of 10 h (112, 113). The corresponding figure for seawater is about 27 ng ml^{-1} (112). DNA can be extracted from such environments to produce fragments of about 10 000 bp that are suitable for restriction analysis. Highest concentrations of dissolved DNA are found in the environments with the highest bacterial biomass and activity. The relatively rapid turnover of extracellular DNA in freshwater suggests that sequences released from GEMs would not survive for long periods unless adsorbed to sediments (113). However, most of the target nucleic acids from samples will be located within intact living or dead bacteria. Generally bacterial cells that have been lysed enzymatically and with detergents release DNA of higher molecular weight than those broken by physical means such as sonication or grinding with alumina or glass powder (93). Particular problems may be encountered with enzymatic lysis for those bacterial species that produce large amounts of extracellular polysaccharide slime since the slime layer reduces access by the enzymes. Detergents can be used to lyse a wide variety of bacteria. However, some detergents quench fluorescence of fluorochromes such as DAPI which are used to quantify DNA (25). Problems of quenching can be overcome by disrupting cells using either sonication alone (98) or in combination with lysozyme treatment (25). Clearly, ideal lysis regimes will vary between different bacterial species so that no method can guarantee the direct extraction of nucleic acids in representative proportions from environmental samples. Consequently lysis protocols tend to be compromises that include most of the features necessary to disrupt commonly isolated bacteria (see, for example, Ref. 62). Non-culturable organisms may remain unlysed because insufficient knowledge is available concerning their cell wall structure.

Two basic approaches have been adopted for recovering nucleic acids from environmental samples. One is a cell extraction procedure in which the bacterial cells are first separated from the environmental matrix by differential centrifugation, then lysed and subjected to differential caesium chloride (CsCl) centrifugation to recover the nucleic acids (62, 156). A large proportion of bacteria do, however, still remain attached to particulates even following extensive extraction procedures and density-gradient centrifugations (10). Humic acids in soils and sediments interfere with many enzymatic reactions, including restriction endonuclease cleavage and the polymerase chain reaction (PCR). Therefore it is generally necessary to use very effective nucleic acid purification procedures. Sodium ascorbate is frequently used as a reducing agent to prevent oxidation of phenols. Phenolic compounds, such as humic acids, may be removed from nucleic acid preparations by adsorption to the insoluble polymer polyvinylpolypyrrolidone (PVPP) (62, 147). The alternative approach to recovering DNA from the environment is to extract nucleic acids directly by physical disruption of cells without separation from the environmental matrix, followed by such procedures as alkaline extraction, ethanol extraction, CsCl centrifugation and hydroxylapatite chromatography (25, 107). The direct extraction approach is generally more time-consuming than the cell extraction procedure, but has been rendered considerably shorter by the substitution of polyethylene glycol precipitation for ethanol-precipitation and the use of PVPP to reduce contamination by humic material (147). The direct lysis method produces more than 10 times higher yields of DNA than the cell extraction procedure (147). However, it appears from [3H]thymidine incorporation experiments that the DNA recovered by cell extraction comes primarily from metabolically active cells whereas direct extraction leads to the isolation of large amounts of DNA from lysed cells or from inside non-active cells. Nucleic acids have been successfully isolated directly from a variety of habitats, including soils (31, 146, 155), sediments (62, 107) and water (45, 141).

Amplification of nucleic acids

Although nucleic acids can be readily extracted from environmental samples, the numbers of target molecules present may be extremely small and below the level needed for detection by conventional hybridization techniques. For example, a single unique marker gene for a GEM may be present at only one or two copies per cell. Thus set against a background of 10^7–10^{11} non-target cells, detection technologies are likely to be severely strained. However, many recent advances have been made in developing methods for the amplification of nucleic acid hybridization signals or the *in vitro* replication of target sequences (38, 87). The amplification of either signal or target not only goes a long way towards solving the problem of very low amounts of target sequences, but also opens up the possibility of 'genetic' analysis of non-culturable organisms, which was hitherto precluded by an inability to acquire sufficient pure genetic material.

Use of the polymerase chain reaction (PCR)   The polymerase chain reaction (PCR) is the most widely used method for the *in vitro* amplification of nucleic acids (83, 124). Starting with a DNA (or RNA) template and appropriate flanking oligonucleotide primers it is possible to amplify target sequences by sequential rounds of polymerization using a thermostable DNA polymerase (or reverse transcriptase) and heat denaturation (Fig. 2.5). The primers normally have different sequences that are complementary to opposite strands of template DNA and flank the segment of sequence to be amplified (typically 180–500 base pairs, but can be longer). Generally the primers are between 20 and 27 nucleotides long (T_m 55–75°C). Template DNA is first heat denatured in the presence of each of the four nucleoside triphosphates and a molar excess of the two amplification primers. The reaction mixture is then cooled to a temperature that permits the primers to anneal to their complementary sequences on the template and the 3′ ends of each primer are extended using the thermostable DNA polymerase from *Thermus aquaticus* (*Taq* polymerase). The cycle of denaturation, annealing and DNA synthesis is then repeated 20 to 25 times with the products of one cycle of amplification acting as templates for the next. The principal product of the reaction is double-stranded DNA whose termini are defined by the sequence of the amplification primers. Amplification levels of about 10^6 are attained following a series of about 25 cycles, after which time the amount of *Taq* polymerase becomes limited. Further amplification can be achieved by diluting the reaction mixture by 10^{-3} to 10^{-4}-fold and using the DNA to drive a new polymerase chain reaction. Following 60 cycles of amplification it is therefore possible to achieve amplification levels of 10^{10}, which allow detection of a single copy of the target sequence even in the presence of a 10^{13}-fold excess of non-target DNA (83, 123).

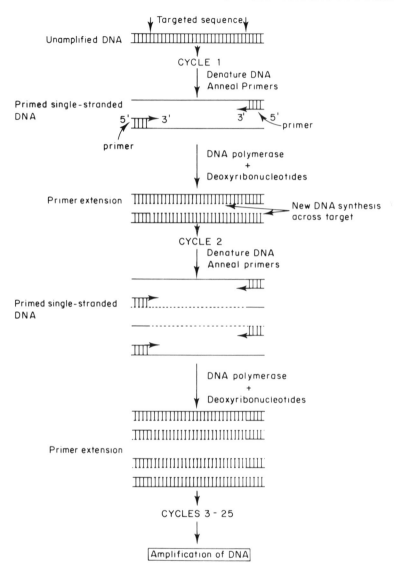

Fig. 2.5 Polymerase chain reaction. Each successive cycle of primer-directed polymerization results in a doubling of the amount of target DNA.

Target sequences in DNA or RNA recovered from environmental samples can be amplified using primers that flank diagnostic sequences for which specific hybridization probes are available. One elegant system that exploits this technique involves a unique sequence marker in the form of a 0.3 kbp *Eco*RI genomic fragment from the napier grass *Pennisetum purpureum* incorporated into a recombinant plasmid carried by *E. coli* (27). This sequence does not occur naturally in prokaryotes and hence will not hybridize to DNA extracted from natural bacterial communities. The presence of GEMs carrying this recombinant can be detected following *in vitro* amplification of the *P. purpureum* sequences using PCR, even when culturable cells of the release host can no longer be detected (27). PCR can be used to allow detection of DNA from as few as one eukaryotic or prokaryotic cell (15, 83). For *E. coli* the detection limits are currently equivalent to those achievable with culture techniques. Thus, as few as one to five viable cells could be detected in 100-ml water samples using PCR amplification of the *E. coli*-specific indicator genes *lacZ* and *lamB* (15). As little as 1–10 fg of *E. coli* DNA could be detected using a *lacZ* probe and primer annealing at 40°C. Higher annealing conditions of 70°C result in less sensitive detection levels down to 100 fg (15). PCR amplification of target sequences from environmental samples gives at least a 10^3-fold increase in sensitivity of detection when compared with non-amplified target DNA using dot blot hybridization (106, 146, 174). Furthermore, as few as 10^2 cells of *Pseudomonas cepacia* could be detected against a background of 10^{11} non-target organisms when using PCR amplification and a *P. cepacia*-specific DNA probe. The sensitivity of detection with PCR can be as low as one target cell per gram of sediment or soil (145, 146). This compares very favourably with conventional colony hybridization using whole gene (10^4 cell ml^{-1}) or rRNA (10^3 cell ml^{-1}) probes. DNA obtained by PCR amplification is also of sufficient quality to be cloned or to be sequenced directly. PCR amplification of specific sequences is also possible *in situ* in animal cells, which opens up the possibility of performing amplifications directly on environmental samples. By coupling single-stranded DNA sequences to magnetic beads it is also possible to recover specific single-stranded complementary sequences from a complex mixture of nucleic acids by using strong magnets as a prelude to PCR amplification (161).

Although PCR is a powerful technique for analysing genes in communities it is not without problems. The yield and efficiency of the process can be estimated crudely from the number of cycles and the amount of input target sequence (124, 146). However, quantitative amplification may be hampered by difficulties in cell lysis and by impurities in environmental samples that may inhibit the polymerization reaction. Furthermore, PCR is inherently highly sensitive to the quality and concentration of reagents used, the denaturation and priming conditions and the rate of temperature change between cycles (125). Care must be taken in the selection of primers to ensure that the correct sequence is amplified, as instances of the unintentional amplification of non-target sequences are not uncommon. Moreover the characteristics of the priming oligonucleotides, notably the formation of self-priming hairpin loops or complementarities between the ends of the two primers, may lead to the amplification of non-target DNA. More accurate quantification may, however, be possible through the use of internal control amplifications using a sequence that is distinguishable from the target but which is amplified using the same primers (125; Fig. 2.6). The enzymatic

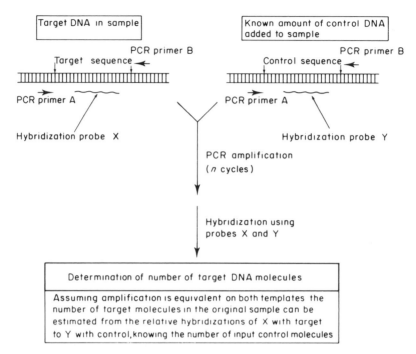

Fig. 2.6 Quantitative amplification of DNA by polymerase chain reaction. Both target and control are flanked by the same sequences to which PCR oligonucleotide primers A and B hybridize and prime DNA synthesis. X and Y are distinct oligonucleotide probes that are uniquely complementary to the target and control sequences respectively.

extension of primer templates is also prone to erroneous incorporation of nucleotides during *in vitro* replication. The maximum error rate reported for *Taq* polymerase following 30 amplification cycles is one misincorporation per 400 bases (123). Provided that sufficient sequence information is obtained from independently isolated clones, however, this has not proved to be a major problem. For practical purposes errors can be ignored if a large number of starting template molecules are used (76). There may, however, be an error rate of about 1% when amplifying a single locus/sequence from a single cell (76). This presents a potential source of error to be encountered when attempting to detect specific GEMs against a large background of other organisms in natural environments. The problem of misincorporation may, however, be overcome by direct sequencing of PCR-amplified fragments since all random nucleotide changes are averaged out (37).

Nucleic acid hybridization technology

Nucleic acid hybridization probes, which commonly comprise specific genes, parts of genes or synthetic oligonucleotides, can be used in appropriate assays to diagnose the presence of particular bacteria in, for example, pathological specimens or environmental samples and to identify genetic differences among isolates of the same species (see, for example, Refs. 12, 51). Such probes may be applied to detect the presence of specific nucleic acid sequences directly in environmental samples as well as in cells obtained following enrichment culture of samples. Specific probes can be directed against the recombinant gene, introduced marker or host DNA/RNA sequence. The use of whole-gene probes based on relatively poorly defined restriction fragments of genomic DNA presents major limitations. These include the fact that the sequences and functions of the DNA concerned may not be known, that the target sequences will normally be present only once on the genome of an organism and that the reaction kinetics of hybridization require long periods of annealing and washing. Short synthetic oligonucleotide probes allow more precise prediction of hybridization conditions, have more favourable reaction kinetics and may be directed more specifically to regions of the target gene or genome (164).

DNA hybridization depends on the reannealing of two complementary nucleic acid strands. Hybridization reactions between target sequences and a probe that is either labelled radioactively or non-isotopically may be carried out either in solution or in solid phase. Solution hybridization permits the screening of large amounts of total DNA from an environmental sample (146). After hybridization has taken place the amount of double-stranded hybrid formed is measured following removal of unannealed probe and target. Between 100 and 1000 cells of a specific bacterium can be detected per gram of environmental sample using this approach (146). Solid-phase hybridization is performed following immobilization of target DNA as single strands, generally on to a nitrocellulose or nylon membrane. *In situ* hybridization is possible provided the target nucleic acid can be accessed in tissue samples or environmental samples such as cores. Following a period of hybridization with the labelled probe, any non-annealed probe can be removed and the amount of hybrid formed is measured by autoradiography for radiolabelled probes, or enzymatically, fluorometrically or by luminescence for non-isotopically labelled probes (84). Non-isotopic methods are becoming more attractive tools for hybridization since they are faster, more convenient and safer than conventional radioactive methods. Colony hybridization, where bacterial cells are grown, transferred to a nitrocellulose filter and lysed directly, provides a method for screening many thousands of culturable bacterial isolates (55). More time-consuming but revealing applications of solid-phase hybridization include Southern (142) transfer hybridization, where DNA restriction fragments, that have been separated on the basis of size on agarose or acrylamide gels, are transferred by blotting on to membranes and identified by hybridization to specific probes. This permits detection of restriction fragment length polymorphisms (RFLPs) that may be characteristic for a particular strain or species.

Comparative sequence analysis of proteins and nucleic acids, particularly of the ribosomal RNA (rRNA) species, has proved to be of enormous value in inferring the natural relationships among microorganisms (108). The presence of conserved sequence domains within rRNA genes has enabled the use of rRNA probes to identify RFLPs of diagnostic significance in the genomes of target organisms (see, for example, Ref. 2). Comparative sequencing of rRNAs has also allowed the synthesis of oligonucleotide hybridization probes for the phylogenetically based identification of bacteria in both pure culture and directly in the environment (49,110). 16S and 5S rRNA molecules have been exploited in such studies. Certain regions of 16S rRNA molecules are conserved among all bacterial species and have therefore proved useful as universal annealing sites for sequencing primers. Other regions are unique to particular genera or species. Care must be taken in sample preparation since endogenous nuclease activity can degrade the RNA resulting in reductions in hybridization signals of up to a thousand-fold (57). Furthermore, by using gene cloning and PCR amplification it has proved possible to identify new bacterial species and their phylogenetic relations even amongst the non-culturable fraction of a number of microbial communities (4, 48, 49, 166). The natural abundance of ribosomes in bacterial cells means that ribosomal RNA species are highly amplified with respect to their DNA coding sequences. For example, in rapidly growing *E. coli* the ribosomal RNA may account for as much as 20% of the dry weight of the cells (110), with each cell containing about 2×10^4 ribosomes in exponential growth. However, it should be noted that the starvation conditions encountered in many natural environments would be expected to lead to a reduction in both the DNA and RNA content of cells with a concomitant reduction in target molecules per cell (105). Moreover, the number of rRNA gene copies is highly variable between bacteria (108,169). The high number of rRNA molecules per bacterial cell, however, normally results in a sensitivity of hybridization of at least 100 times that achieved with DNA targets (41). A further advantage that may be exploited is that RNA-RNA hybrids are more stable than DNA-RNA or DNA-DNA hybrids, which allows the use of more stringent hybridization conditions and hence results in greater specificity and sensitivity of detection (67).

Identification of single cells is possible using fluorescent oligonucleotide probes labelled, for example, with tetramethylrhodamine and homologous to species-determinative regions of 16S rRNA (3–5, 35). These probes can be used as stains for epifluorescence microscopy to analyse microorganisms fixed on films, since such cells are permeable to short oligonucleotides (35). Moreover, by combining such probes with flow cytometry (see later section, p. 45, and Chapter 1) it is possible to analyse mixed populations of microorganisms in suspension.

Nucleic acid sequencing for species identification

Although automated techniques are now available that permit the sequencing of entire genomes for identification purposes, analysis of rRNA sequences is providing a more convenient and rapid approach to the problem of identifying and tracking GEMs. The natural amplification of rRNA means that these molecules can be readily purified from cultures or samples and sequenced (78, 118).

Initial studies using 5S rRNA have proved to be useful in analysing bacterial populations, but are limited by the relatively small number of nucleotides involved (about 150) which restricts discrimination between species (132, 144). More information and discrimination can be obtained by examining the sequences of 16S rRNA (approximately 1500 nucleotides) or 23S rRNA (approximately 2300 nucleotides) (78). Rapid genotypic identification and classification of bacteria is also possible, without sequencing, by analysis of transfer RNA (tRNA) and 5S rRNA profiles on high resolution polyacrylamide gels. The patterns obtained reflect genotypic relationships determined by sequencing 16S and 23S RNA species (61). 16S rRNA sequences may be obtained from natural isolates either by sequencing the rRNA genes concerned (108, 110), or more effectively, by extracting the rRNA and using it as a template for reverse transcriptase with appropriate oligonucleotide primers that anneal to conserved regions to produce complementary DNA (cDNA) which can subsequently be cloned and sequenced (166, 169). Alternatively, the DNA coding sequence for rRNA may be amplified from bulk DNA isolated from environmental samples, using PCR techniques and primers that hybridize to conserved regions, followed by cloning and sequencing (28, 37, 49). Thus by careful choice of primers it is possible to achieve group-, species- or even strain-specific recovery of 16S or 23S rRNA sequences from environmental samples.

The ability to acquire such information directly by cloning rRNA sequences is analogous to the culture of bacteria followed by phenotypic characterization. The power of rRNA techniques resides not only in their ability to detect low levels of a particular organism, but also in the fact that they can be used to determine both the presence and genetic diversity of non-culturable bacteria (49, 166). Thus it is now possible to begin to understand the genetic structure of natural bacterial communities and hence how the deliberate or accidental introduction of novel organisms might perturb habitats and their microbial populations. Furthermore, the introduction of a site-specific mutation or oligonucleotide signature into a non-essential region of an rRNA gene could provide a useful, naturally amplified marker system for GEMs (141).

Immunological methods of detection

Immunofluorescence microscopy, in which bacteria may be labelled using specific antibodies conjugated to a fluorochrome, can permit the direct observation of selected strains or species (20, 22, 54). This method has been used to identify specific bacteria in coal waste (7), soil (20) and the marine environment (165). Immunofluorescence microscopy can only be used to detect a GEM if antibodies are available that can react with a specific gene product that is both accessible on the cell surface and diagnostic for the released organism. Unless the target antigen is chosen carefully there may be cross-reaction with structurally identified components on the surface of unrelated microorganisms. Furthermore, the reacting antigens would have to be expressed consistently in all cells of the released population. Monoclonal antibodies against *Flavobacterium* P25 have been shown to be able to detect as few as 20 target bacteria per gram of soil even under starvation conditions (94). However, results will depend largely on the specific nature of the antigen to which the antibodies are raised and the degree to which its expression occurs in natural habitats. Also environmental conditions may, for example, induce the production of capsular polysaccharides in the target bacteria, which may obscure surface-located antigens that are accessible under laboratory culture conditions. Problems are also encountered in soil due to interference by particulate matter and the natural autofluorescence of some soil components. Nevertheless, immunofluorescence reactions may be used in conjunction with direct viable counts and/or vital dyes in order to distinguish living bacteria from dead cells or particulate matter.

Immunofluorescence may also be used in conjunction with flow cytometry. Flow cytometry involves the use of finely focused laser light beams that permit the counting, characterization and sorting of individual cells. Immunofluorescence flow cytometry has a number of advantages over immunofluorescence microscopy, not least of which are increased sensitivity and the potential for automation (115). Furthermore, provided two or more different antibodies are available that can be labelled with different coloured fluorochromes it is possible to separate and count various members of mixed populations. Fluorescent stains are also available for DNA, RNA and other components and oligonucleotide probes may be labelled fluorescently (4, 5). This means that nucleic acid content and other specific properties of labelled cells can be examined and cell populations sorted electronically on the basis of their fluorescence at different wavelengths. However, since the cells must be in suspension in an aqueous environment bacteria cannot be examined *in situ* in soil or other solid-phase environments and must be extracted beforehand. Nevertheless, this method is of great potential use, particularly in the observation of planktonic populations of bacteria in aquatic environments.

Immunological specificity may also be used to achieve selective enrichment or recovery of specific bacteria. For example, immunomagnetic capture of recombinant pseudomonads was made possible by coupling a monoclonal antibody that is specific to the flagella of a particular strain of *Pseudomonas putida* to the surface of magnetized polystyrene beads (103, 126). Target release bacteria expressing the specific flagellar antigen become attached when the beads are suspended in lakewater or other aqueous samples. The bead-bacteria complexes can then be recovered using a strong magnet and washed to remove non-specifically-bound bacteria (Fig. 2.7). The success of this procedure depends largely on the specificity of the antibodies and the number of washes necessary to remove non-target bacteria. Flagella are found on a wide variety of bacteria and are sufficiently variable in amino acid sequence in some antigenic domains to allow production of strain-specific monoclonal antibodies suitable for use in immunomagnetic capture. These

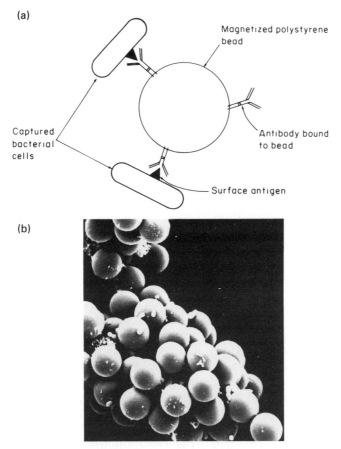

Fig. 2.7 Immunomagnetic capture of recombinant bacteria. (a) Diagrammatic representation of immunomagnetic capture using antibody-coated polystyrene beads. Antibody-coated beads are added to the environmental sample. Following antibody–antigen reaction the bead–bacterial complexes are recovered by using strong magnets and washed to remove any non-specifically adhered bacteria. (b) Scanning electron micrograph of recombinant *Pseudomonas putida* captured from lakewater sample using polystyrene beads coated with *P. putida*-specific antibody.

surface structures are, however, rather brittle in nature and may reduce the efficiency of the capture process. An advantage, nevertheless, is that captured viable bacteria can be removed easily from the beads for subsequent culture or other analysis simply by subjecting the bead-bacteria complexes to vigorous vortexing (103). The flagella antigens seem to be expressed on the surface of *P. putida* released into freshwater-sediment microcosms for as long as the organisms survive (see Chapter 3). The potential problem of the relative fragility of the capture antigen is, however, not encountered with surface antigens that are integral to the microbial surface. For example, spores of a genetically manipulated *Streptomyces lividans* strain have been recovered from soil samples using magnetized beads coated with a mono-clonal antibody to a spore coat antigen. Recovery efficiencies varied depending on the relative number of spores to beads and the nature of the soil matrix (175). In view of the difficulties of extracting bacteria from natural solid-state environments and the large numbers of washing steps needed to produce absolute specificity it is likely that the immunomagnetic capture strategy will be most effective for either the recovery of high numbers of target cells, with the acceptance of some contaminating particles, or the inefficient recovery of a very small number of target cells.

GENE TRANSFER

Genetic exchange occurs naturally among bacterial populations with transfer efficiencies varying greatly within and between species, depending on individual propensities for transferring genes and on prevailing environmental conditions. The fitness and competitiveness of an introduced GEM will influence its ability to transfer genes to the environmental pool: the longer the organism survives, the greater the chance of colonizing a habitat and of transferring genetic material at a detectable frequency. If transferred genes become established within indigenous bacteria, further genetic exchanges may occur, in turn enlarging the arena for genetic interplay. Moreover, genetic rearrangements may arise in microbial populations and these might be likely to occur more frequently in a natural ecosystem than in a laboratory setting since many of the processes, e.g. mutation, transposition and recombination, that would be instrumental in effecting such changes appear to be induced by environmental stresses (77, 100, 131, 163). The GEM may itself be subjected to genetic rearrangement, may receive new genes from the natural microflora and may acquire an altered genotype. The techniques and marker systems devised should thus be capable of accommodating the various genetic interactions that may occur. Three natural gene transfer mechanisms are recognized in bacteria, namely transformation, transduction and conjugation.

Transformation

Transformation involves a complex and varying series of biochemical events that lead to the uptake, functional maintenance and expression of naked DNA by the host (127, 149). Although such extracellular DNA is vulnerable to degradation by nucleases in the environment its adsorption to clay particles and other surfaces appears to decrease susceptibility. Furthermore, binding of nucleases to such particles may reduce nucleolytic activity and hence the degradation of the DNA. Indeed, concentrations of DNA sufficient to support transformation have been reported in various habitats (113, 114; see also earlier section, p. 37).

For naturally transformable microorganisms competency may require the accumulation of a critical concentration of competence factor (CF) in the growth medium, as is the case with streptococci (80, 97, 153). This may be difficult to achieve in environments with a low solid:water ratio (since this would cause considerable dilution of CF), and/or that contain copious quantities of proteases. However, adsorption of CF to clay minerals may reduce degradation by proteases. In other bacteria, for example *Haemophilus* species, competence can be induced by a period of unbalanced growth, effected for example, by a shift-down in nutrient status (140, 150). Unbalanced growth conditions are probably the norm in natural habitats and particularly in the soil. Thus such organisms would normally be in a state of competence for transformation unless they are spore-forming species that exist predominantly as spores in nature. In many transformation systems competent bacteria are capable of taking up both homospecific and heterospecific DNA. However, *Haemophilus*, *Campylobacter* and *Neisseria* species show a preference for DNA from the same or closely related species. Specific recognition sequences appear to be involved in directing DNA uptake in these species (50, 130, 139). Heterologous DNA lacking such a sequence can sometimes be taken up with varying degrees of efficiency. Bacterial transformation has been described in soil inoculated with strains of *Bacillus subtilis* (52, 53).

Transduction

Transduction appears to be responsible for genetic transfer in the environment. The process has been reported in a number of bacteria, including *Pseudomonas aeruginosa* (104, 129), *Vibrio parahaemolyticus* (13) and *E. coli* (177, 178). Successful transduction relies on a coordinated series of events, that involves:

1. The successful infection of a host bacterium.
2. Incorporation of bacterial genes into progeny virions.
3. Release of such virions into the environment where other susceptible bacteria are available to elicit further infections.
4. Introduction and establishment of the transduced bacterial DNA in the recipient.
5. Repeated spread to other bacteria in the habitat.

It is not yet clear how variation in the physical, chemical and biological features of the natural environment may perturb such a series of events. Significant titres of both virulent and temperate phages have been found in aquatic and terrestrial environments (47, 143). Whilst free phages in the environment can be inactivated and may rapidly suffer loss of infectivity, survival can be enhanced by their adsorption to particulates (9, 86). Packaging of DNA, introduced in a GEM, into a phage might thus protect it from degradation for longer than it otherwise would if housed in its microbial carrier or if released from moribund or dead cells. Significantly the DNA might persist in the virion phenotypically undetected, until it could infect a susceptible host. Lysogenic bacteria seem to be effective recipients in transduction (129). This is probably a consequence of the transduced recipient being immune to infection by superinfecting phages. Superinfection immunity would thus be expected to increase the likelihood of the successful establishment of transduced cells in the environment. Furthermore, the pool of transducing phages could be replenished by induction of the prophages from lysogenic bacteria. This process, which appears to be stimulated by a number of environmental stresses (8, 74, 75, 163), would thereby ensure that the potential for transductional exchange is constantly maintained. However, the host range of bacteriophages is generally limited due to the requirement for appropriate cell surface receptors. Hence transduction is unlikely to be an important contributor to interspecies or intergeneric transfer. Yet such transfers are a major consideration when assessing the potential hazardous consequences of released genes.

Conjugation

By far the most likely route for dissemination of genes among and between bacterial populations in the environment is conjugation. Some conjugative elements can transfer to a diversity of bacterial genera and even to fungi and plant cells (see, for example, Refs 58, 59, 96, 159, 160, 176). Conjugative plasmids can mobilize for transfer non-conjugative plasmids and chromosomal genes. Furthermore, conjugative plasmids may have a wider conjugal transfer host range than replication host range (21, 152). Consequently such plasmids can deliver non-conjugative plasmids and transposable elements into cells in which the conjugative plasmids cannot themselves establish. Such features of the conjugation process indicate that it may be relatively non-specific and mediate the opportunistic expansion of the host range of genes. Conjugative transfer has been demonstrated in a variety of habitats, including lakes (109), rivers (11), soils (32, 119, 162, 168, 170), on plant surfaces (39, 72), in sewage treatment facilities (46, 91, 99) as well as in humans and animals (81, 151).

Factors infuencing the dispersal of genes

A myriad of physical, chemical and biological factors can influence the transfer, establishment and expression of genetic information. Competitive pressures exerted by the native microflora of a particular habitat are likely to reduce establishment of the introduced strain(s) and influence gene transfer. It would be expected *a priori* that transfer frequencies measured from donor A to recipient B would be much lower in natural mixed populations than in the artificial milieu of a laboratory bipartite cross involving the mixture of two monocultures. Indeed, transfer has been observed at reduced frequencies in the presence of natural microflora in various environments (see, for example, Refs 11, 90, 168).

A variety of different measurements have been made in order to determine the efficiency of plasmid transfer in laboratory and environmental conditions. Transfer may be measured by determining the relative number of transconjugants formed as a function of time (6, 167), the ratio of transconjugants to donor (34) or recipient cells (11) after a fixed time, or by measurement of the rate constant of transfer (138). The problem with comparing transfer efficiencies in natural populations is that there is no standard method for expression and even simple parameters, such as number of input donors and recipients, are often not stated in ecological studies (128). The ability of plasmids to mediate the formation of mating aggregates, comprising different proportions of donor and recipient cells (92), can also complicate the estimation of transfer kinetics. Furthermore, the dynamics of plasmid transfer are more complex when bacteria are growing on surfaces or in colonies rather than in liquid suspension (137). Transitory derepression of conjugative functions can affect transfer through natural populations. This phenomenon, which is manifested in newly formed transconjugants, permits temporary derepression of plasmid-encoded transfer functions that would otherwise be inhibited by a repressor (88). In this way rates of donation of plasmids can transiently be increased. There is some evidence that such derepression occurs particularly strongly when bacteria are growing on surfaces (137).

Successful establishment of DNA that has been transported to recipients by whichever route demands recombination and/or replication depending on the nature of the transferred material. In the case of chromosomal genes homologous recombination with the resident genome is required. This process is, by its very nature, conservative and may serve to limit the flow of heterologous DNA in the environment. However, in *E. coli* the extent of DNA homology for successful recombination may be as little as 23–27 base pairs (136). Such lengths of sequence similarity can be displayed by very diverse species of bacteria. Furthermore, recombination events may occur more readily in the environment than in the laboratory, since expression of *recA*, the cardinal gene for homologous recombination of *E. coli*, *Pseudomonas aeruginosa* and other bacteria, is inducible by various stresses likely to be encountered in an environmental setting. Incompatibility between introduced and resident replicons can also affect the establishment of DNA within recipients (16) and pose a barrier to the spread of genetic material in the environment. However, incompatibility can sometimes be circumvented by recombination between the replicons involved.

A further important element in natural transfer is the DNA methylation (modification) and restriction rates of the primary and intermediate donors in a mixed population (126). Restricting hosts will obviously reduce the ingress of particular genes if they arrive from a host lacking the appropriate modification methylase specificity. However, exposure of the host to periods of stress, such as elevated temperature or UV, may transiently eliminate restriction and so permit establishment of transferred genes (163). In recent years several restriction systems in *E. coli* that only restrict DNA that is already methylated at specific sites have been discovered (120). Moreover, many bacterial species seem to methylate their DNA for reasons other than protection against self-restriction. For example, the *dam* methylase (56) of *E. coli* is involved in recognition of newly synthesized daughter DNA strands in the process of mismatch repair and in regulation of some genes. Such methylations may interfere with restriction endonuclease cleavage at the relevant sites. Thus a gene or plasmid passing through a network of strains may gain and/or lose methylation specificities that may prove advantageous or disadvantageous depending upon the restriction system operating. In this context, the use of restriction-less recipients in transfer experiments provides a system for measuring donor potential and a base line for 'worse case scenarios' for escape of a gene from a GEM by any of the known transfer mechanisms.

Evidence from *in vivo* and *in situ* studies thus indicates that horizontal gene transfer can occur in a diversity of natural habitats and with a range of different hosts, but is likely to be sporadic. A repertoire of transfer mechanisms is potentially available. However, in the natural environment logistics must play an essential part in initiating transfer. The concentration of the components of a particular transfer system and the probability of such components encountering each other in the same ecosystem are fundamental to the occurrence of any transfer event. Of the various cellular barriers that may be operating to impede the traffic of genes and reduce successful establishment of genetic information, none will probably be absolute. Furthermore, alternative stategies may be available to surmount certain barriers. Thus if the population density of an introduced GEM is sufficiently high and the recombinant DNA persists for long enough, the potential exists for transfer to the environmental gene pool and for the generation of genetic novelty. Even if a very low transfer frequency prevails, a gene that confers a selective advantage on its host in a particular environmental niche will be encouraged to survive, establish and spread. This highlights a need for detection technologies that are sensitive enough to track rare genetic events. However, further data on the exchange of genetic information between GEMs and the autochthonous microbiota and on the effects of environmental factors on such exchanges are necessary before the contribution of gene transfer to any potentially hazardous consequence of releasing a GEM into the environment can be critically assessed.

FUTURE PROSPECTS

A variety of molecular genetic and immunological techniques is now being brought to bear on the problem of monitoring the fate of GEMs in the natural environment. Particular difficulties will be experienced due to the sheer magnitude of sampling of the environments into which GEMs may be released. Consequently it is desirable that detection technologies are capable of automation so that large numbers of samples can be processed in a short period of time. The advent of DNA amplification technologies together with an increasing knowledge of the discriminating properties of rRNA sequences make automated nucleic acid extraction and sequencing techniques a tenable proposition for the monitoring of microbial communities. Recent advances in the detection of antigen-antibody complex formation, by such techniques as surface plasmon resonance (SPR) (33), are likely to speed up processes of immunological screening. What is already clear is that successful monitoring of the fate and activity of GEMs in the open environment is unlikely to be achieved by the use of a single marker gene or detection technique and that combinations will be necessary. Moreover, unpredictability of the interaction of genes within the environment will mean that every candidate release organism will have to be considered with respect to its individual properties.

REFERENCES

1. Al-Hadithi, S.A. & Goulder, R. (1989) *Letters in Applied Microbiology* **8**, 87–90.
2. Altwegg, M., Hickman-Brenner, F.W. & Farmer, J.J. (1989) *Journal of Infectious Diseases* **160**, 145–149.
3. Amann, R.I., Krumholz, L. & Stahl, D.A. (1990) *Journal of Bacteriology* **172**, 762–770.
4. Amann, R.I., Binder, B.J., Olsen, R.J., Chisholm, S.W., Deveraux, R. & Stahl, D.A. (1990) *Applied and Environmental Microbiology* **56**, 1919–1925.
5. Amann, R.I., Stromley, J., Deveraux, R., Key, R. & Stahl, D.A. (1992) *Applied and Environmental Microbiology* **58**, 614–623.
6. Anderson, E.S. (1975) *Nature* **255**, 502–504.
7. Apel, W.A., Dugan, P.R., Filippi, J.A. & Rheins, M.A. (1976) *Applied and Environmental Microbiology* **32**, 159–165.
8. Armentrout, R.W. & Rutberg, L. (1971) *Journal of Virology* **8**, 455–468.
9. Babich, H. & Stotzky, G. (1980) *Water Research* **14**, 185–187.
10. Bakken, L.R. (1985) *Applied and Environmental Microbiology* **49**, 1482–1487.
11. Bale, M.J., Fry, J.C. & Day, M.J. (1987) *Journal of General Microbiology* **133**, 3099–3107.
12. Barkay, T., Fouts, D.L. & Olson, B.H. (1985) *Applied and Environmental Microbiology* **49**, 486–490.
13. Baross, J.H., Liston, J. & Morita, R.Y. (1974) In *International Symposium on Vibrio parahaemolyticus*, Eds Fujino, T., Sakaguchi, G., Sakazaki, R. & Takeda, Y. Saikon Publishing, Tokyo, pp. 129–137.
14. Bej, A.K., Perlin, M.H. & Atlas, R.M. (1988) *Applied and Environmental Microbiology* **54**, 2472–2477.
15. Bej, A.S., Steffan, R.J., DiCesare, J., Haff, L. & Atlas, R.M. (1990) *Applied and Environmental Microbiology* **56**, 307–314.
16. Bergquist, P.L. (1987) In *Plasmids—A Practical Approach*, Ed. Hardy, K. IRL Press, Oxford, pp. 37–78.
17. Bergstrom, I., Heianen, A. & Salonen, K. (1986) *Applied and Environmental Microbiology* **51**, 664–667.
18. Bishop, D.H.L., Entwistle, P.F., Cameron, I.R., Allen, C.J. & Possee, R.D. (1988) In *The Release of Genetically-Engineered Microorganisms*, Eds Sussman, M., Collins, C.H., Skinner, F.A. & Stewart-Tull, D.E. Academic Press, London, pp. 143–179.
19. Bochner, B.R. (1989) *Nature* **339**, 157–158.
20. Bolhool, B.B. & Schmidt, E.L. (1980) *Advances in Microbial Ecology* **4**, 203–241.

21. Boulnois, G.J., Varley, J.M., Sharpe, G.S. & Franklin, F.C.H. (1985) *Molecular and General Genetics* **200**, 65–67.
22. Brayton, P.R., Tamplin, M.L., Huq, A. & Colwell, R.R. (1987) *Applied and Environmental Microbiology* **54**, 2862–2865.
23. Britten, R.J. & Kohne, D.E. (1968) *Science* **161**, 529–540.
24. Brunt, C.F., Jones, K.C. & James, T.W. (1979) *Analytical Biochemistry* **92**, 497–500.
25. Burrison, B.K. & Nuttley, D.J. (1990) *Applied and Environmental Microbiology* **56**, 362–365.
26. Byrd, J.J. & Colwell, R.R. (1990) *Applied and Environmental Microbiology* **56**, 2104–2107.
27. Chaudry, G.R., Toranzos, G.A. & Bhatti, A.R. (1989) *Applied and Environmental Microbiology* **55**, 1301–1304.
28. Chen, K., Neimark, H., Rumore, P. & Steinman, C.R. (1989) *FEMS Microbiology Letters* **57**, 19–24.
29. Colwell, R.R. (1987) *Journal of Applied Bacteriology* (Symposium Supplement) 1S–6S.
30. Colwell, R.R., Brayton, P.R., Grimes, D.J., Roszak, D.B., Huq, S.A. & Palmer, L.M. (1985) *Bio/Technology* **3**, 817–820.
31. Cresswell, N., Saunders, V.A. & Wellington, E.M.H. (1991) *Letters in Applied Microbiology* **13**, 193–197.
32. Cresswell, N., Herron, P.R., Saunders, V.A. & Wellington, E.M.H. (1992) *Journal of General Microbiology* **138**, 659–666.
33. Cullen, D.C., Brown, R.G.W. & Lowe, C.R. (1988) *Biosensors* **3**, 211–225.
34. Curtiss, R., Caro, L.G., Allison, D.P. & Stallions, D.R. (1969) *Journal of Bacteriology* **100**, 1091–1104.
35. DeLong, E.F., Wickham, G.S. & Pace, N.R. (1989) *Science* **243**, 1360–1363.
36. Drahos, D.J., Hemming, B.C. & McPherson, S. (1986) *BioTechnology* **4**, 439–444.
37. Edwards, U., Rogall, T., Blocker, H., Emde, M. & Bottger, E.C. (1989) *Nucleic Acids Research* **17**, 7843–7853.
38. Fahrlander, P.D. (1988) *Bio/Technology* **6**, 1165–1168.
39. Farrand, S.K. (1989) In *Gene Transfer in the Environment*, Eds Levy, S.B. & Miller, R.V. McGraw-Hill, New York, pp. 261–285.
40. Ford, S.F. & Olson, B. (1988) *Advances in Microbial Ecology* **10**, 45–79.
41. Forsman, M., Sandstrom, G. & Jaurin, B. (1990) *Applied and Environmental Microbiology* **56**, 949–955.
42. Francisco, D.E., Mah, R.A. & Rabin, A.C. (1973) *Transactions of the American Microscopical Society* **92**, 416–421.
43. Fry, J.C. & Zia, T. (1982) *Journal of Applied Bacteriology* **53**, 189–198.
44. Fuhrman, J.A. & Azam, F. (1982) *Marine Biology* **66**, 109–120.
45. Fuhrman, J.A., Comeau, D.E., Hagstrom, A. & Chan, A.M. (1988) *Applied and Environmental Microbiology* **54**, 1426–1429.
46. Gealt, M.A. (1989) In *Guide to Environmental Microbiology*, Eds Levin, M., Seidler, R. & Pritchard, P.H. American Society for Microbiology, Washington DC.
47. Germida, J.J. & Khachatourians, G.G. (1988) *Canadian Journal of Microbiology* **34**, 190–193.
48. Giovannoni, S.J., Britschgi, T.B., Moyer, C.L. & Field, K.G. (1990) *Nature* **345**, 60–63.

49. Giovannoni, S.J., DeLong, E.F., Olsen, G.J. & Pace, N.R. (1988) *Journal of Bacteriology* **170**, 720–726.
50. Goodgal, S.H. (1982) *Annual Review of Genetics* **16**, 169–192.
51. Goodwin, P.H., Kirkpatrick, B.C. & Duniway, J.M. (1990) *Applied and Environmental Microbiology* **56**, 669–674.
52. Graham, J.B. & Istock, C.A. (1978) *Molecular and General Genetics* **166**, 287–290.
53. Graham, J.B. & Istock, C.A. (1979) *Science* **204**, 637–639.
54. Hall, G.H., Jones, J.G., Pickup, R.W. & Simon, B. (1990) *Methods in Microbiology* **23**, 181–210.
55. Hanahan, D. & Meselson, M. (1980) *Gene* **10**, 63–67.
56. Hattman, S., Brooks, J.E. & Maswekar, M. (1978) *Journal of Molecular Biology* **126**, 367–380.
57. Haun, G. & Gobel, U. (1987) *FEMS Microbiology Letters* **43**, 187–193.
58. Heinemann, J.A. (1991) *Trends in Genetics* **7**, 181–185.
59. Heinemann, J.A. & Sprague, G.F. (1989) *Nature* **340**, 205–209.
60. Hobbie, J.E., Daley, R.J. & Jasper, S. (1977) *Applied and Environmental Microbiology* **33**, 1225–1228.
61. Hofle, M.G. (1990) *Archives of Microbiology* **153**, 299–304.
62. Holben, W.E., Jansson, J.K., Chelm, B.K. & Tiedje, J.M. (1988) *Applied and Environmental Microbiology* **54**, 703–711.
63. Hussong, D., Colwell, R.R., O'Brien, M., Weiss, E., Pearson, A.D., Weiner, R.M. & Burge, W.D. (1987) *Bio/Technology* **5**, 947–950.
64. Iriberri, J., Unanue, M., Ayo, B., Bararia, I. & Egea, L. (1990) *Applied and Environmental Microbiology* **56**, 483–487.
65. Jannasch, H.W. (1969) *Journal of Bacteriology* **99**, 156–160.
66. Jefferson. R.A. (1989) *Nature* **342**, 837–838.
67. Jiang, X., Estes, M.K. & Metcalf, T.G. (1987) *Applied and Environmental Microbiology* **46**, 2487–2495.
68. Jones, J.G. (1987) In *Ecology of Microbial Communities*, Eds Fletcher, M., Gray, T.R. & Jones, J.G. *Society for General Microbiology Symposium* **41**, 235–260.
69. Jones, J.G. & Simon, B.M. (1975) *Journal of Applied Bacteriology* **39**, 317–329.
70. Jones, J.G., Gardener, S., Simon, B.M. & Pickup, R.W. (1986) *Journal of Applied Bacteriology* **60**, 455–462.
71. Knight, I.T., Shults, S., Kaspar, C.W. & Colwell, R.R. (1990) *Applied and Environmental Microbiology* **56**, 1059–1066.
72. Knudsen, G.R., Walter, M.V., Porteous, L.A., Prince, V.J., Armstrong, J.L. & Seidler, R.J. (1988) *Applied and Environmental Microbiology* **54**, 343–347.
73. Kogure, K., Simidu, U. & Taga, N. (1979) *Canadian Journal of Microbiology* **25**, 415–420.
74. Kokjohn, T.A. & Miller, R.V. (1987) *Journal of Bacteriology* **169**, 1499–1509.
75. Kokjohn, T.A. & Miller, R.V. (1988) *Journal of Bacteriology* **170**, 578–582.
76. Krawczak, M., Reiss, J., Schmidtke, J. & Rosler, U. (1989) *Nucleic Acids Research* **17**, 2197–2201.
77. Kretschmer, P.J. & Cohen, S.N. (1979) *Journal of Bacteriology* **139**, 515–519.
78. Lane, D.J., Pace, B., Olsen, G.J., Stahl, D.A., Sogin, M.L. & Pace, N.R. (1985) *Proceedings of the National Academy of Sciencies of the USA* **82**, 6955–6959.

79. Lee, S. & Fuhrman, J.A. (1990) *Applied and Environmental Microbiology* **56**, 739–746.
80. Leonard, C.G. & Cole, R.M. (1972) *Journal of Bacteriology* **110**, 273–280.
81. Levy, S.B. & Marshall, B.M. (1988) In *The Release of Genetically-Engineered Microorganisms*, Eds Sussman, M., Collin, C.H., Skinner, F.A. & Stuart-Tull, D.E. Academic Press, London, pp. 61–76.
82. Levy, S.B. & Miller, R.V. (Eds) (1989) *Gene Transfer in the Environment.* McGraw-Hill, New York.
83. Li, H., Gyllensten, U.B., Cui, X., Saiki, R.K., Erlich, H.A. & Arnheim, N. (1988) *Nature* **335**, 414–417.
84. Lichter, P. & Ward, D.C. (1990) *Nature* **345**, 93–94.
85. Lindow, S.E. & Panopoulos, N.J. (1988) In *The Release of Genetically-Engineered Microorganisms*, Eds. Sussman, M., Collins, C.H., Skinner, F.A. & Stewart-Tull, D.E. Academic Press, London, pp. 121–138.
86. Lipson, S.M. & Stotzky, G. (1987) In *Human Viruses in Sediments, Sludges and Soils*, Eds Rao, V.C. & Melnick, J.L. CRC Press, Boca Raton, pp. 198–229.
87. Lizardi, P.M., Guerra, C.E., Lomeli, H., Tussie-Luna, I. & Kramer, F.R. (1988) *Bio/Technology* **6**, 1197–1202.
88. Lundquist, P.D. & Levin, B.R. (1986) *Genetics* **113**, 483–497.
89. MacNaughton, S.J., Rose, D.A. & O'Donnell, A.G. (1992) *Journal of General Microbiology* **138**, 667–673.
90. Manceau, C., Gardon, L. & Devaux, M. (1986) *Canadian Journal of Microbiology* **32**, 835–841.
91. Mancini, P., Ferteis, S., Nave, D. & Gealt, M.A. (1987) *Applied and Environmental Microbiology* **53**, 665–671.
92. Manning, P.A. & Achtman, M. (1980) In *Bacterial Outer Membranes. Biogenesis and Functions*, Ed. Inouye, M. John Wiley, New York, pp. 406–447.
93. Marmur, J. (1963) *Methods of Enzymology* **6**, 726–738.
94. Mason, J. & Burns, R. (1990) *FEMS Microbiology Letters* **73**, 299–308.
95. Matsuyama, T. (1984) *FEMS Microbiology Letters* **21**, 153–157.
96. Mazodier, P., Petter, R. & Thompson, C. (1989) *Journal of Bacteriology* **171**, 3583–3585.
97. McCarty, M. (1980) *Annual Reviews of Genetics* **14**, 1–15.
98. McCoy, W.F. & Olson, B.H. (1985) *Applied and Environmental Microbiology* **49**, 811–817.
99. McPherson, P. & Gealt, M.A. (1986) *Applied and Environmental Microbiology* **51**, 904–909.
100. Miller, R.V. & Kokjohn, T.A. (1988) *Journal of Bacteriology* **170**, 2385–2387.
101. Molin, S., Klemm, P., Poulsen, L.K., Biehl, H., Gerdes, K. & Andersson, P. (1987) *Bio/Technology* **5**, 1315–1318.
102. Morgan, J.A.W., Winstanley, C., Pickup, R.W., Jones, J.G. & Saunders, J.R. (1989) *Applied and Environmental Microbiology* **55**, 2537–2546.
103. Morgan, J.A.W., Winstanley, C., Pickup, R.W. & Saunders, J.R. (1991) *Applied and Environmental Microbiology* **57**, 503–509.
104. Morrison, W.D., Miller, R.V. & Sayler, G.S. (1978) *Applied and Environmental Microbiology* **36**, 724–730.
105. Moyer, C.L. & Morita, R.Y. (1989) *Applied and Environmental Microbiology* **55**, 2710–2716.
106. Neilson, J.W., Josephson, K.L., Pillai, S.D. & Pepper, I.L. (1992) *Applied and Environmental Microbiology* **58**, 1271–1275.

107. Ogram, A., Sayler, G.S. & Barkay, T. (1987) *Journal of Microbiological Methods* **7**, 57–66.
108. Olsen, G.J., Lane, D.L., Giovannoni, S.J., Pace, N.R. & Stahl, D.A. (1986) *Annual Review of Microbiology* **40**, 337–365.
109. O'Morchoe, S., Ogunseitan, O., Sayler, G.S. & Miller, R.V. (1988) *Applied and Environmental Microbiology* **54**, 1923–1929.
110. Pace, N.R., Stahl, D.A., Lane, D.J. & Olsen, G.J. (1986) *Advances in Microbial Ecology* **9**, 1–55.
111. Paul, J.H. & David, A.W. (1989) *Applied and Environmental Microbiology* **55**, 1865–1869.
112. Paul, J.H. & Myers, B. (1982) *Applied and Environmental Microbiology* **43**, 1393–1399.
113. Paul, J.H., Jeffrey, W.H., David, A.W., DeFlaun, M.F. & Cazares, L.H. (1989) *Applied and Environmental Microbiology* **55**, 1823–1828.
114. Paul, J.H., Jeffrey, W.H. & Defleur, M.E. (1987) *Applied and Environmental Microbiology* **53**, 170–179.
115. Phillips, A.P. & Martin, K.L. (1988) *Journal of Immunological Methods* **106**, 109–117.
116. Pickup, R.W. (1991) *Journal of General Microbiology* **137**, 1009–1019.
117. Porter, K.G. & Feig, Y.B. (1980) *Limnology and Oceanography* **25**, 943–948.
118. Qu, L.H., Michot, B. & Bachellerie, J.-P. (1983) *Nucleic Acids Research* **11**, 5903–5920.
119. Rafii, F. & Crawford, D.L. (1988) *Applied and Environmental Microbiology* **54**, 1334–1340.
120. Raleigh, E.A. & Wilson, G. (1986) *Proceedings of the National Academy of Sciences of the USA* **83**, 9070–9074.
121. Rattray, E.A., Prosser, J.I., Killhorn, K. & Glover, L.A. (1990) *Applied and Environmental Microbiology* **56**, 3368–3374.
122. Ross, P., O'Gara, F. & Condon, S. (1990) *Applied and Environmental Microbiology* **56**, 2164–2169.
123. Saiki, R.K., Gelfand, D.H., Stoffel, S., Scharf, S.J., Higuchi, H., Horn, G.T., Mullis, K.B. & Erlich, H.A. (1988) *Science* **239**, 487–494.
124. Saiki, R.K., Scharf, S., Faloona, F., Mullis, K.B., Horn, G.T., Erlich, H.A. & Arnheim, N. (1985) *Science* **230**, 1350–1354.
125. Sambrook, J., Fritsch, E.F. & Maniatis, T. (1989) *Molecular Cloning: a Laboratory Manual*. Cold Spring Harbor Laboratory, Cold Spring Harbor.
126. Saunders, J.R., Morgan, J.A.W., Winstanley, C., Raitt, F.C., Carter, J.P., Pickup, R.W., Jones, J.G. & Saunders, V.A. (1990) In *Bacterial genetics in the environment*, Eds Fry, J. & Day, M. Chapman & Hall, London, pp. 3–21.
127. Saunders, J.R. & Saunders, V.A. (1988) *Methods in Microbiology* **21**, 79–128.
128. Saunders, J.R. & Saunders, V.A. (1992) In *Genetic Interactions Between Microorganisms in the Microenvironment*, Eds Wellington, E.M. & Van Elsas, J.D. Manchester University Press (in press).
129. Saye, D.J., Ogunseitan, O., Sayler, G.S. & Miller, R.V. (1987) *Applied and Environmental Microbiology* **53**, 987–995.
130. Scocca, J.J. (1990) *Molecular Microbiology* **4**, 321–327.
131. Sedgwick, S.G. (1986) In *Accuracy in Molecular Processes, its Control and Relevance to Living Systems*, Eds Kirkwood, T.B.L., Rosenberger, R.F. & Galas, D.J., Chapman & Hall, Cambridge, pp. 233–289.
132. Seewaldt, E. & Stackebrandt, E. (1982) *Nature* **295**, 618–620.

133. Selander, R.K., Caugant, D.A., Ochman, H., Musser, J.M., Gilmour, M.N. & Whittam, T.S. (1986) *Applied and Environmental Microbiology* **51**, 873–884.
134. Shaw, J.J. & Kado, C.I. (1986) *Bio/Technology* **4**, 560–564.
135. Shaw, J.J., Dove, F., Geiger, D. & Kloepper, J.W. (1992) *Applied and Enviromental Microbiology* **58**, 267–273.
136. Shen, P. & Huang, H.V. (1986) *Genetics* **112**, 441–457.
137. Simonsen, L. (1990) *Journal of General Microbiology* **136**, 1001–1007.
138. Simonsen, L., Gordon, D.M., Stewart, F.M. & Levin, B.R. (1990) *Journal of General Microbiology* **136**, 2319–2326.
139. Sisco, K.L. & Smith, H.O. (1979) *Proceedings of the National Academy of Sciences of the USA* **76**, 972–976.
140. Smith, H.O., Danner, D.B. & Deich, R.A. (1981) *Annual Review of Biochemistry* **50**, 41–68.
141. Somerville, C.C., Knight, I.T., Straube, W.L. & Colwell, R.T. (1989) *Applied and Environmental Microbiology* **55**, 548–554.
142. Southern, E.M. (1975) *Journal of Molecular Biology* **98**, 503–517.
143. Soyal, S.M., Gerba, C.P. & Bitton, G. (Eds) (1987) *Phage Ecology*. John Wiley & Sons, London.
144. Stahl, D.A., Lane, D.J., Olsen, G.J. & Pace, N.R. (1985) *Applied and Environmental Microbiology* **49**, 1379–1384.
145. Steffan, R.J. & Atlas, R.M. (1988) *Applied and Environmental Microbiology* **54**, 2185–2191.
146. Steffan, R.J. & Atlas, R.M. (1990) *Biotechniques* **8**, 316–318.
147. Steffan, R.J., Goksoyr, J., Bej, A.K. & Atlas, R.M. (1988) *Applied and Environmental Microbiology* **54**, 2908–2915.
148. Stewart, G.S.A.B. (1989) *Letters in Applied Microbiology* **10**, 1–8.
149. Stewart, G.J. (1989) In *Gene Transfer in the Natural Environment*, Eds Levy, S.B. & Miller, R.V. McGraw-Hill, USA, pp. 139–164.
150. Stewart, G.J. & Carlson, C.A. (1986) *Annual Review of Microbiology* **40**, 211–235.
151. Stotzky, G. & Babich, H. (1986) *Advances in Applied Microbiology* **31**, 93–138.
152. Thomas, C.M. & Smith, C.A. (1987) *Annual Review of Microbiology* **41**, 77–101.
153. Tomasz, A. & Hotchkiss, R.D. (1964) *Proceedings of the National Academy of Sciences of the USA* **51**, 480–487.
154. Torella, F. & Morita, R.Y. (1981) *Applied and Environmental Microbiology* **41**, 518–527.
155. Torsvik, V. (1980) *Soil Biology and Biochemisty* **12**, 15–21.
156. Torsvik, V.L. & Goksoyr, J. (1978) *Soil Biology and Biochemistry* **10**, 7–12.
157. Torsvik, V., Goksoyr, J. & Daae, F.L. (1990) *Applied and Environmental Microbiology* **56**, 782–787.
158. Torsvik, V., Salte, K., Sorheim, R. & Goksoyr, J. (1990) *Applied and Environmental Microbiology* **56**, 776–781.
159. Trieu-Cuot, P., Carlier, C., Martin, P. & Courvalin, P. (1987) *FEMS Microbiology Letters* **48**, 289–294.
160. Trieu-Cuot, P., Carlier, C. & Courvalin, P. (1988) *Journal of Bacteriology* **170**, 4388–4391.
161. Uhlen, M. (1989) *Nature* **340**, 733–734.
162. van Elsas, J.D., Govaert, J.M. & van Veen, J.A. (1987) *Soil Biology and Biochemistry* **19**, 639–647.

163. Walker, G.C. (1984) *Microbiological Reviews* **48**, 60–93.
164. Wallace, B., Shaffer, J., Murphy, R.F., Bonner, J. & Itakura, K. (1979) *Nucleic Acids Research* **11**, 3543–3557.
165. Ward, B.B. & Carlucci, A.F. (1985) *Applied and Environmental Microbiology* **50**, 149–201.
166. Ward, D.M., Weller, R. & Bateson, M.M. (1990) *Nature* **345**, 63–65.
167. Watanabe, T. (1963) *Bacteriological Reviews* **27**, 87–115.
168. Weinberg, S.R. & Stotzky, G. (1972) *Soil Biology and Biochemistry* **4**, 171–180.
169. Weller, R. & Ward, D.M. (1989) *Applied and Environmental Microbiology* **55**, 1818–1822.
170. Wellington, E.M.H., Cresswell, N. & Saunders, V.A. (1990) *Applied and Environmental Microbiology* **56**, 1413–1419.
171. Winstanley, C., Morgan, J.A.W., Pickup, R.W., Jones, J.G. & Saunders, J.R. (1989) *Applied and Environmental Microbiology* **55**, 771–777.
172. Winstanley, C., Morgan, J.A.W., Pickup, R.W. & Saunders, J.R. (1991) *Applied and Environmental Microbiology* **57**, 1905–1913.
173. Winstanley, C., Morgan, J.A.W., Pickup, R.W. & Saunders, J.R. (1992) In *Genetic Interactions Between Microorganisms in the Microenvironment*, Eds Wellington, E.M. & van Elsas, J.D. Manchester University Press (in press).
174. Wipat, A., Wellington, E.M.H. & Saunders, V.A. (1991) *Applied and Environmental Microbiology* **57**, 3322–3330.
175. Wipat, A., Wellington, E.M. & Saunders, V.A. (1992) In *Genetic Interactions Between Microorganisms in the Microenvironment*, Eds. Wellington, E.M. & van Elsas, J.D. Manchester University Press (in press).
176. Zambryski, P., Tempe, J. & Schell, J. (1989) *Cell* **56**, 193–201.
177. Zeph, L.R., Onaga, M.M. & Stotzky, G. (1988) *Applied and Environmental Microbiology* **54**, 1731–1737.
178. Zeph, L.R. & Stotzky, G. (1989) *Applied and Environmental Microbiology* **55**, 661–665.

Chapter 3

In Situ *Detection of Plasmid Transfer in the Aquatic Environment*

R. W. Pickup[1]*, J. A. W. Morgan[2] and C. Winstanley[3]
[1]Institute of Freshwater Ecology, Windermere Laboratory, Ambleside
[2]Freshwater Biological Association, Windermere Laboratory, Ambleside
[3]Department of Genetics & Microbiology, University of Liverpool

INTRODUCTION

The release of genetically engineered mircoorganisms (GEMs) into the environment is a reality brought about by technologies that enable microorganisms to be altered genetically. Depending on the nature of their genetic modifications, GEMs can be classified into two groups. Firstly, some will be designed to be cultured in contained industrial facilities where their commercial by-product is extracted on-line before final processing and purification. Industrially important microorganisms have been successfully exploited in this way for the production of a range of pharmaceutical products such as interferon, insulin and human growth hormone (13,118). Secondly, an increasing number of microorganisms are being specifically developed for

---

* Corresponding author.

Monitoring Genetically Manipulated Microorganisms in the Environment. Edited by C. Edwards
Published 1993 John Wiley & Sons Ltd. © 1993 R.W. Pickup, J.A.W. Morgan and C. Winstanley

release at high concentrations into the environment where containment may not be effective. The benefits for releasing such organisms are not in doubt. They have the potential to become powerful tools for environmental management purposes (111). Microorganisms can be adapted or engineered to degrade a range of toxic compounds which accumulate due to their refractory nature. Detoxification of polluted areas such as land-fill or sites where accidental spillages have occurred is an attractive possibility (85). The benefits of releasing engineered microbes for agricultural purposes include improved crop yield, an enhancement of crops' ability to fix atmospheric nitrogen, pest control that obviates the need for chemical biocides, and protection of crops against adverse weather conditions such as frost damage (106).

The potential benefits of such biotechnological progress must be balanced by the realization that there is a lack of information on the effects of releasing high concentrations of novel microorganisms into the environment. It is likely that most deliberate releases into soil for agricultural purposes, or accidental spillages from a contained laboratory or biotechnology plant, will ultimately lead to some GEMs becoming dispersed through run-off water systems into lakes and streams (9) or directly from the land to groundwater aquifers. As a result of our inability to contain the GEMs totally within the target site, their introduction into the environment, either deliberately or accidentally, is not without potential risks. Concern has focused on the effects that the release of GEMs would have on the environment. This follows the release of animal and plant species which have become pests in the environment (52, 96). It is possible that the genetic modifications present within engineered organisms may enhance their survival and increase the risk of adverse ecological effects. These may include disruptive effects on biotic communities; the introduction of engineered microorganisms could alter the balance among species and lead to changes in community level characteristics such as diversity or the elimination of pivotal role species through competition or interference (59). Similarly, introduction of GEMs could adversely affect ecosystem function by interfering with vital natural processes such as nutrient cycling and energy flow. Neuhold & Ruggiero (73) identified several process and community dysfunctions which could have an adverse affect on the ecosystem. Allied to this is the possibility that recombinant DNA may be directly transmitted to other indigenous species through conjugation, transformation and transduction processes (104).

One concern is that microorganisms, through genetic modification or through the receipt of recombinant DNA in the laboratory or in the environment, will become pathogenic. Abelson (1) used a worst case model in which an *Escherichia coli* host carrying polyoma DNA (DNA from viruses that induce tumours) was found to induce tumours in animals. However, the possibility of unintentionally creating a serious pathogen by engineering a safe microorganism appears remote when considered in the context of what is know about the molecular basis of disease (14, 100). One or several introduced genes are unlikely to transform a bacterium into a pathogen.

Usually several specific genes working interactively are required by bacteria to cause an infectious disease (100). Although pathogenicity is an important factor in risk assessment, it is likely to be governed by the release strain rather than the recombinant DNA it carries. It is unlikely that recombinant DNA encoding animal or crop plant pathogenicity will be introduced into bacteria destined for environmental purposes. Rigorous testing of release candidates will avoid the problem of knowingly introducing a microorganism that is pathogenic to humans, animals or plants.

It is important that the potential for gene transfer is considered when assessing the risks associated with environmental release. There is little or no information on the transfer of recombinant DNA in the environment. This is partly due to restrictions preventing release into the environment for purely experimental purposes placed by the regulatory authorities. As a consequence most experimentation is laboratory-based and in reality may not relate to the potential for such events to occur in nature. Therefore, we must rely on the information gained through more conventional studies (for review see Ref. 27). This chapter is directed towards assessing the available information on plasmid transfer obtained or extrapolated from *in situ* studies in the aquatic environment and how this relates to the transfer potential of introduced microorganisms.

GENE TRANSFER IN THE AQUATIC ENVIRONMENT

DNA transfer in bacteria has been shown to occur by three distinct routes: transformation, transduction and conjugation (61). Although this chapter will deal almost exclusively with plasmid transfer events assumed to have occurred through conjugal processes, the contribution of transformation and transduction should not be underestimated. There is increasing evidence that the latter two processes may play a greater role in gene transfer *per se* (71). As a result the boundaries which distinguish the trichotomy of transfer processes have become less defined.

Transformation is the process by which free DNA can be taken into the cell. The presence of extracellular DNA has been demonstrated in freshwater and seawater (20, 21, 78). A number of bacterial strains have been shown to release or produce extracellular DNA in aquatic environments (75). In addition, members of the natural population have been observed to incorporate this free DNA. However, this DNA could have been taken up as gene sequences or individual bases. In freshwater, DNA turnover is very rapid (77) and recombinant DNA sequences from GEMs would therefore not be expected to survive for long periods in the environment. Although attachment of DNA to particulate material may improve its survival, its persistence is still very short (4–24 hours; Ref. 78). Protection of DNA in excreted vesicles outside the cell could improve the chance for gene exchange in the environment; such structures have been reported in strains of *Neisseria gonorrhoeae* and *Haemophilus* sp. (22, 51). Transformation has not actually been demonstrated *in situ* in any aquatic environment although its potential to occur remains high (102).

Transduction is the transfer of bacterial genes by bacteriophages. Since transfer would have to occur between host populations subjected to infection by the same phage strain, the narrow host-range of most phages reduces the likelihood of gene transfer within a mixed bacterial population. Chromosomal DNA has been shown to be transduced in cells released into environmental test chambers incubated in freshwater. Plasmid DNA has also been transduced in a similar system (91). In the presence of a natural microflora the transfer frequency was reduced, possibly due to the decline of the release host. Since phage are released in a free form and do not require cell-to-cell contact, transduction may represent an ideal method for dispersing genes in the environment. Similarly, Amin & Day (4) demonstratd F116-mediated chromosomal transduction between *Pseudomonas aeruginosa* strains on submerged river stones. There is also evidence that the adhesion of bacteriophages to particles can offer protection against inactivation and prolong their persistence in the environment (5, 23). The gene transfer potential of phages has been further emphasized by the observation of high numbers of viruses found in the marine and possibly the freshwater environments (12,86).

Conjugation is the transfer of plasmid DNA through direct cell-to-cell contact and was considered the most likely route for plasmid transfer in the environment (87). The molecular basis for conjugal transfer has been subjected to many excellent reviews (see, for example, Ref. 117).

THE AQUATIC ENVIRONMENT AND ITS EFFECT ON PLASMID TRANSFER

The aquatic environment comprises a wide diversity of habitats. These include the water column itself and a range of habitats associated with the littoral and benthic sediments and those associated with a range of particle sizes and surfaces provided by plants and rocks. It has been proposed that conjugal transfer is dependent on relative concentration of the donor and recipient (58). Bacterial numbers vary between different habitats and between sites within the same habitat. For example, the viable count in the water column for a series of freshwater lakes in the English Lake District was approximately 10^3 ml^{-1}. However, the bacterial viable count for Windermere and Ennerdale showed a 200-fold difference although the direct counts were comparable to each other at 10^5 to 10^6 (47). The presence of sediments substantially increases the bacterial viable and direct counts (approximately x 10^6 and 10^{10} cells g^{-1} dry weight respectively; Ref. 42). The presence of organic films is known to concentrate bacteria in the epilithon encapsulating them under a protective layer of mucilage. Such assemblages in sub-arctic freshwater streams were found to have cell densities of 0.5–20 x 10^7 cm^{-2} although, as with any direct counting method, not all were considered to be metabolically active (62). Therefore the concentration of bacteria found in different habitats is highly variable. Bacterial numbers also vary significantly with season (42). The factors responsible for this variation are numerous. They include geographic location, underlying geology, light penetration, temperature, pH, dissolved oxygen and nutrient concentration (55,63). Many of these factors have a profound effect on the physiological state of indigenous microorganisms and, indirectly, on their transfer potential (89). For example, microbial competition and predation (2, 98) not only affect the donor and recipient strains but will greatly influence the survival, establishment and proliferation of any potential transconjugants. However, in the presence of both abiotic and biotic pressures (the most extreme being the absence of nutrients), a plasmid-containing *E. coli* isolated from the environment was able to transfer its plasmids (25).

Gene transfer frequency can be directly affected by a variety of biological factors relating to the types of genes undergoing transfer and the genetic status of the donor and recipient cells (64, 71: see Chapter 2). These include plasmid incompatibility, host range, DNA restriction and modification, transmissibility and gene expression.

EVIDENCE FOR PLASMIDS IN AQUATIC BACTERIA

Plasmids encoding a wide range of functions are ubiquitous in the large variety of organisms isolated from diverse environments (101). Many of these traits confer selective advantages to the host cell in response to stressed or hostile environments. Of particular importance in the aquatic environment are plasmids that offer protection against UV damage (88), heavy metal pollution (39, 95), or xenobiotic pollution (36). However, not all plasmids can be linked to a particular function, yet their ubiquity remains (35, 80). The term 'cryptic' has been assigned to plasmids to which no function has been attributed. It does not imply that they are functionless and that they confer no selective advantage. Independent investigations into the occurrence of plasmids (Table 3.1) revealed that 9–75% of heterotrophic bacteria from diverse aquatic environments ranging from the water/air interface through the water column to the sediments carried one or more plasmids. These data have revealed that plasmids in aquatic bacteria ranged in size from 3–>400 kb. Considering that 15–30 kb is thought to be the minimum size for a transmissible plasmid (117) and that those below this size may be mobilizable by other plasmids then the potential for conjugal transfer is high. Failure to detect plasmids in some isolates does not necessarily imply their absence. It is possible that their plasmids are refractory to purification by certain methods; for example, some large TOL plasmids could not be isolated using well established methods involving caesium chloride purification (e.g. Ref. 38) but could be isolated by a more subtle method which yielded DNA of sufficient purity for restriction analysis and cloning (79, 115). Although a variety of plasmid isolation methods are available no single technique can be considered to be universal (110).

Table 3.1 Occurrence of plasmids in bacteria from aquatic habitats.

Bacterial type	% Plasmid-containing bacteria	Size (kb)	Location	Reference
E. coli	70	nd	River	91
Vibrio sp.	35	nd	Marine	32
Heterotroph	46	3->60	Estuary	27
Heterotroph	9–15	50–400	River sediment	16
Heterotroph	43		Marine	87
Pseudomonas	73	5–100	Marine amphipod	109
Heterotroph	nd	5–200	Epilithon	19
Heterotroph	70	30->350	River	72
Heterotroph	33	1–100	Lake	85
Psychrophile	42	>15	Sediment	52
Psychrophile	20	>15	Seawater	52
Heterotroph	33	nd	Marine sediment	24
Heterotroph	45	nd	Water/air interface	36

nd = not determined; kb = kilobase pairs.

EVIDENCE FOR RECIPIENTS IN THE AQUATIC ENVIRONMENT

There are numerous reports on the ability of laboratory-developed strains to act as donors and recipients in plasmid transfer experiments under laboratory conditions. Walter and co-workers (114) compared the standard laboratory transfer techniques (broth mating, membrane filter mating, colony cross streak mating and combined spread mating) for their sensitivity in assessing the transfer of recombinant DNA into a range of indigenous bacteria. They concluded that since no single technique was completely reliable in detecting transconjugants then one which integrated all the transfer procedures was necessary. Using the combined method, recombinant plasmids were successfully transferred to indigenous bacteria isolated from plants and soil. Schilf & Klingmuller (93) demonstrated that 18% of the bacteria isolated from water possessed the capacity to act as recipients for plasmids under laboratory and simulated environmental conditions. A range of bacteria isolated from the aquatic environment were tested for their ability to receive a broad-host-range plasmid (R68) and to receive a non-conjugative plasmid (R1162) through mobilization by R68 (30). Of the 68 isolates tested 38% successfully received the plasmid R68 and 15% acquired R1162 through R68-mediated mobilization. Although the isolates could not be considered completely representative of the aquatic community the results showed that 42% of the pseudomonads tested and 38% of the unidentified Gram-negative rods (representing the viable and culturable bacterial population) were able to act as recipients for these plasmids. The potential for the viable but non-culturable population, which in some habitats represents >98% of the total direct count (46), remains to be determined.

METHODS FOR ASSESSING PLASMID TRANSFER IN THE ENVIRONMENT

In situ methodology

Plasmid transfer in the aquatic environment has been studied for many years. Initial concern was expressed over the release from sewage and waste treatment plants of antibiotic-resistant bacteria carrying plasmids, and the effect that *in situ* transfer of these plasmids would have on the indigenous microbial flora and on human health (105, 111). Early experiments were laboratory oriented and focused on R-plasmid transfer between *E. coli* and coliforms (41, 101). The first *in situ* experiments involved mixing donors (carrying a known plasmid) with a defined recipient strain. The mating mixtures were incubated at environmental temperatures in dialysis tubing placed in lake or river water (34). The enclosed membrane system for *in situ* transfer experiments has progressed from dialysis tubing (33) to Teflon film bags (91). The membrane diffusion chamber used by McFeters & Stuart (70) for monitoring the survival and viability of faecal indicator bacteria was modified for plasmid transfer studies (3, 64, 74). The advantage of using enclosed membrane systems lies in the ability to make direct comparisons between experiments performed in the field and those performed in the laboratory. Enclosed membrane systems also prevent the organisms from being released into the environment, yet permit nutrients and ions to enter. This allows the organisms to experience the prevailing environmental conditions in the aquatic system (112).

Bale *et al.* (6, 7) developed a novel *in situ* method to investigate unenclosed plasmid transfer in the epilithon of a Welsh river. Filters containing both donor and recipient cells were strapped to the surface of sterile scrubbed stones and placed in an open-topped beaker and submerged in the riverwater. The filters were then removed after a period of time and transported under sterile conditions to the laboratory where the bacterial growth was removed by scrubbing. The suspension was analysed for the presence of the donors and recipients, and transconjugants were counted using selective media (7). This system was also used in the laboratory where various environmental parameters could be strictly controlled (7). Rochelle *et al.* (88) modified the procedure by examining the transfer potential of epilithic bacteria by scrubbing the stones to remove the indigenous epilithic bacteria producing a mixed natural suspension (MNS). This MNS was plated directly onto selective media and the plasmid content of the selected bacteria analysed; those containing detectable plasmids could then undergo *in vitro* analysis for conjugal plasmids. Alternatively, the MNS was mixed on filters with a suitable recipient and transconjugants were selected as an indication that conjugal plasmids occur in the epilithic population (88). These experiments with MNS represent a bridge between *in situ* and *in vitro* experiments (28).

The use of microcosms in the analysis of plasmid transfer

The use of microcosms also represents an acceptable compromise between *in situ* experimentation and plasmid transfer analysis by more traditional laboratory-orientated methods (112). A microcosm is a controlled, reproducible laboratory system designed to contain the necessary components of the ecosystem which is being investigated. Since natural systems are complex, microcosms introduce a degree of simplification which may exclude or alter some of the processes which occur in natural ecosystems (112). Microcosms can be classified in two ways. Firstly, there are those which are meant to represent a subunit of nature, transported into the laboratory, with native organisms and biological processes relatively undisturbed. Secondly, there are microcosms which are simpler in design, perhaps constructed in the laboratory, which allow certain biological principles to be investigated (84). A series of recommendations were made regarding the design of microcosms at the First International Conference on the Release of Genetically Engineered Micro-organisms. The recommendations included (a) that each microcosm should be clearly defined, (b) that results from any microcosm should be interpreted with a full understanding of the limitations of that system and (c) that the microcosm should be calibrated with field conditions (84).

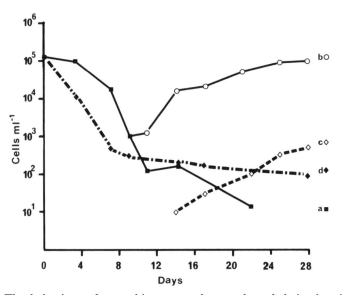

Fig. 3.1 The behaviour of recombinant pseudomonads and their plasmids in lake water and sterile lakewater. *Pseudomonas putida*, PaW340 containing plasmid pLV1010 declined in numbers after release into sterile lakewater at a concentration of 10^5 cells ml^{-1} (a). Plasmid-free segregants appeared after 8 days and re-established the initial inoculum concentration (b). Cells containing pLV1010 which had undergone a large deletion emerged after 14 days (c). *Pseudomonas putida*, PaW340 declined in numbers and eventually became undetectable after 28 days following release into lakewater (d). No segregation was observed.

The experimental systems based on dialysis tubing or membrane diffusion chambers have been described as laboratory or field microcosms, as can continuous culture methods (112). It is unlikely that gene transfer will take place in the water column due to a number of reasons but most importantly cell concentrations and lack of solid surfaces. Therefore, microcosms have been developed which represent aquatic sediment systems which are analogous to terrestrial soil microcosms except that they are tri-phasic, comprising air, water and sediment. Development of these microcosms has been for purposes other than gene transfer experiments such as those investigating bacterial establishment and survival and the toxicological effects of pollutants on biological processes (2, 72, 107). However, experiences with such systems will lead to their more effective use in plasmid transfer experiments. Sterile lakewater microcosms, comprising filtered then autoclaved lakewater in sterilized conical flasks, were used to study the survival of recombinant pseudomonads carrying conjugative and non-conjugative marker plasmids (72, 81). Although not representative of natural water bodies, the lakewater microcosms were useful for examining the behaviour of the host organism and its plasmid under non-selective and non-competitive conditions (Fig. 3.1). The behaviour of the recombinant pseudomonads was monitored under the same conditions in microcosms comprising non-treated lakewater (72). The differences in behaviour are presented in Fig. 3.1. In lakewater, where the released organism is subject to predation and competition, survival is severely reduced and, accordingly, so is its transfer potential (Fig. 3.1d).

Sediment microcosms can be developed in a variety of contained systems. Sediment from the aquatic environment can be obtained as a slurry using mechanical grabs or as an intact sediment core using a coring instrument such as a Jenkin sampler (37). The latter instrument recovers intact and representative sediment cores (Fig. 3.2) which can be used as a model microcosm. Their size also permits simultaneous replicate experiments to be performed. Survival experiments have also been performed with sediment mini-cores (Fig. 3.3) which comprise sediment slurry in a 30-ml syringe tube which is sealed with stopper but open to the air through a syringe needle. Survival of recombinant *E. coli* strains in lake sediment was similar to the pseudomonads in lakewater although more prolonged (Pickup, R.W. *et al*, unpublished data; Fig. 3.4). Sediments are equally manageable if deposited in covered beakers with overlying lakewater. Increasing the size of microcosms may reduce the number of replicate experiments which can be performed at any one time but may produce a system more representative of the environment under investigation. An artificial stream was constructed in glass containers 1-m in length (Pickup, unpublished data; Fig. 3.5) which comprised lake sediment fed continuously with water directly from a natural water source. The versatility of this system is that many such units can be operated simultaneously either in series or in parallel. The sampling of sediments by non-coring methods destroys preformed nutrient concentration gradients. However, in the artificial stream, mini-core and beaker sediment, gradient conditions rapidly re-established to produce a representative sediment upon

Overlying
lakewater

Sediment (13 cm)

Fig. 3.2 Intact core of lake sediment with overlying water obtained using a Jenkin sampler.

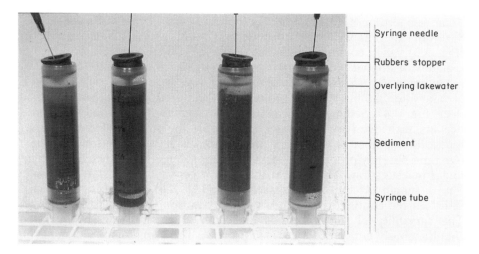

Syringe needle

Rubbers stopper

Overlying lakewater

Sediment

Syringe tube

Fig. 3.3 Example of mini-core microcosms comprising lakewater sediment in 30-ml syringe tube.

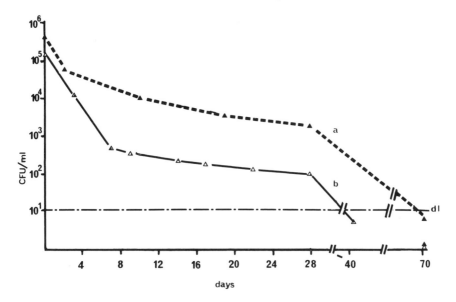

Fig. 3.4   Survival of *E. coli* in sediment (▲) and *Pseudomonas putida* in lakewater (△). *E. coli* was detectable up to 70 days compared to 28 days for *P. putida* in lakewater.

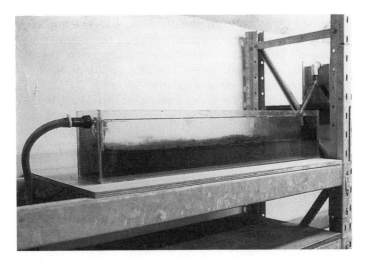

Fig. 3.5   Large-scale continuous flow microcosm simulating a freshwater stream. The system comprises lake sediment with overlying but continually replenished stream water in 1-m glass containers.

settling. Microcosms which simulate groundwater aquifers have been used successfully to show that introduced pseudomonads could be maintained and contribute to the removal of toxic pollutants (43). Microcosm technology and the development of test protocols, despite its inherent limitations, will be a useful if not an essential tool for environmental risk assessment.

Retrospective plasmid analysis

In situ investigations can be complemented by the information gained from the analysis of the epidemiology of plasmids in natural populations (45, 80, 95, 108). Retrospective analysis of plasmid transfer involves the isolation of specific groups of bacteria such as pseudomonads (80), mercury-resistant bacteria (45) or antibiotic-resistant bacteria (108). Plasmids from these bacteria are then extracted by a variety of methods and compared structurally using a range of molecular methods (113). The occurrence of transfer events may then be inferred through the similarities found between plasmid structures isolated from related and non-related bacteria. However, this type of analysis is limited by the efficacy of plasmid extraction techniques (66,110) and the variety of indigenous bacteria which can be isolated through culture methods.

DIRECT EVIDENCE OF PLASMID TRANSFER IN THE AQUATIC ENVIRONMENT

In situ experiments

The data revealed by *in situ* experiments show that the transfer frequency is highly variable, from high (10^{-1} transconjugants per donor cell) to virtually undetectable levels (10^{-9} transconjugants per donor cell). This is not surprising considering the variety of methods and plasmids used and the lack of comparable environmental data available for each habitat. The transfer frequency data cannot therefore be compared directly. The complexity of each ecosystem and the number of environmental variables which have to be considered may preclude any direct, meaningful comparison between transfer frequencies obtained in different habitats. Therefore the reliance on measurement of transfer frequency as a barometer for likely risk to the environment through uncontained dissemination of recombinant genes must be questioned (see Chapter 2).

The value of measuring transfer frequency for comparative studies in the same habitat, with environmental parameters as clearly defined as possible, is not in doubt. The early *in situ* experiments indicated that temperature may influence transfer frequencies (34). In laboratory studies temperature had a clear effect on plasmids, which were usually of clinical origin. Many plasmids such as RP1 show a transfer optimum above 35°C with frequency increasing until lethal temperatures were encountered (53). Rochelle *et al.* (88) compared the behaviour of mercury resistance plasmids captured from epilithic MNS using a laboratory-defined recipient. Although the frequency at which they transferred varied, all exhibited the same temperature profile and optima (approximately 20°C). Conjugal transfer has been shown to occur at temperatures normally found in the aquatic environment. *In situ* transfer was detected in the River Taff between 7°C and 20°C with optimal transfer of mercury resistance at 15–17°C. O'Morchoe *et al.* (74) showed that R68.45 could transfer in diffusion chambers at 16°C in reservoir water. Bale *et al.* (7) demonstrated that the *in situ* transfer frequency of pQM1 followed the environmental temperature very closely throughout the year. Linear regression of the transfer frequency ($\log_{10}$) in relation to the environmental temperature revealed that for every 2.6°C increase in temperature, a 10-fold increase in transfer frequency was detected. When the MNS was used as a source of plasmids, temperature (optimum 25°C), pH and organic nutrient supply affected transfer frequency. Synergistic effects of pH and temperature suggest that combinations of other variables will also have complex effects, thus making prediction of plasmid transfer events in the environment difficult. O'Morchoe *et al.* (74) demonstrated the transfer of the plasmids R68.45 and FP5 in diffusion chambers in a reservoir and in laboratory simulations (see Table 3.2). They noted, as did Bale *et al.* (7), that the transfer frequencies decreased in the presence of indigenous bacteria. No transfer of the plasmids R68.45 and FP5 to the indigenous population was detected.

Table 3.2 *In situ* plasmid transfer in aquatic environments.

Environment	Plasmid	Frequency*	Method‡	Reference
River	R	10^{-3}	D	30
Wastewater	R	10^{-5}	M	59
Sewage	R	10^{-6}	M	3
Pond	RP4	10^{-7}	D	84
Pond	R1	10^{-7}	M	29
Epilithon	pQM1	$10^{-1.9}$	Stone/filter	6
Reservoir	R68.45	$10^{-3.6}†$	M	67
MNS epilithon	Various	$10^{-3.8}†$	Laboratory	79

* Frequency expressed as transconjugants per recipient cell ml^{-1} except † where frequency was expressed as transconjugants per donor cell ml^{-1}. ‡ D = dialysis tubing; M = membrane diffusion chamber.

In situ experiments also suggested that plasmids which are considered to be stably maintained in laboratory culture were destabilized once released into the environment. The frequency of loss of antibiotic resistance markers and loss of transfer function after transfer was 100-fold higher *in situ* than was observed in laboratory simulations (74). Exposure to environmental conditions may increase the frequency of genetic rearrangements. Rochelle *et al.* (88) found that although highly related with respect to restriction endonuclease pattern, some of the pQM series of plasmids obtained from the MNS carried DNA insertions associated with streptomycin resistance. Similarly, a group of plasmids were isolated from a freshwater stream which were shown to be highly related although one was found to have small structural differences (80). The transfer event does not necessarily cause the destabilization. Survival experiments involving *Pseudomonas putida* mt-2 carrying a TOL::RP4 co-integrate plasmid, PAW53–4 (54) showed that in sterile lakewater the plasmid underwent several significant genetic rearrangements (Pickup *et al.*, unpublished data). These included loss of the plasmid, deletion of the TOL genes and retention of the RP4 section, transposition of the TOL genes into the host chromosome and recovery of a transposable element from the host chromosome. Instability of the plasmids was associated with periods of rapid growth in the microcosm.

Waste treatment facilities are aquatic systems which provide ideal conditions for microbial growth and may be conducive to microbial interactions, including gene transfer (92). This chapter will not dwell on the complexities of such facilities suffice to note that many of the bacteria detected contain conjugative plasmids (see Ref. 92); *in situ* plasmid transfer has been demonstrated (see, for example, Refs 3,29,64) as well as mobilization of non-conjugative cloning vectors (65). An example of the interaction which can occur between indigenous and a genetically manipulated bacteria was observed by McClure and co-workers (68). *Pseudomonas putida* UWC1 was released into a simulated activated sludge unit (ASU) and its survival monitored (68). The strain contained a non-conjugative but mobilizable plasmid, pD10, which encoded 3-chlorobenzoate (3CB) degradation. Although UWC1 (pD10) remained detectable for more than 8 weeks it did not enhance the breakdown of 3CB. After 36 days, strains of UCW1 were re-isolated from the ASU which had the ability to transfer the pD10 associated genotype in laboratory conjugation experiments. These isolates were shown to have acquired, since their introduction into the ASU, a range of plasmids which were capable of mobilizing pD10. Furthermore, laboratory filter mating experiments showed that bacteria indigenous to the ASU were able to act as recipients for pD10 and actively expressed the genes carried on that plasmid. However, no direct transfer of pD10 to indigenous bacteria was detected in the ASU (68). These results indicate that the release of recombinant genes on non-conjugative vectors is no guarantee that containment in the original release host will be assured, and that the transfer potential for recombinant DNA may be derived from the indigenous microbial population (65,68). Waste treatment may become a focal point for research as GEMs may indirectly enter these facilities or be engineered and released to improve the degradation of recalcitrant compounds (32).

Retrospective plasmid analysis

Several reports provide information about bacterial plasmids in aquatic bacteria (see Table 3.1). Early studies concentrated on the presence or absence of plasmid DNA in bacterial isolates from river, estuarine and marine environments (see Table 3.1) rather than plasmid structure. Hada & Sizemore (36), Glassman & McNicol (31) and Burton *et al.* (16) showed that higher numbers of plasmids occurred in bacteria isolated from polluted sites compared to those obtained from clean control sites. The plasmid content of the bacteria increased, with those found in polluted sites carrying larger plasmids and in some cases multiple plasmids (16, 31, 116). Where a specific pollutant has been applied the increase in plasmid-containing strains is often complemented by an increase in the number of related strains within that population, each conferring a selective advantage on its host (45, 56, 88). By implication the increase in related plasmids could be associated with plasmid transfer in addition to genetic rearrangements which produce the structural differences. The pollutant stresses also cause shifts in taxonomic diversity. It is possible that bacteria which are unable to receive plasmids which encode a particular selective advantage, and those which fail to adapt to a particular stress, will be lost (8, 99). This loss would increase the proportion of plasmid-containing strains in the population leaving an opportunity for the plasmid-containing cells to proliferate. Therefore, in retrospective studies transfer should only be implied when related plasmids are detected in different species or genera.

In a study of copper-resistant bacteria at several sites along a freshwater stream, related plasmids were detected that did not encode copper resistance or enhance the survival of the bacterial host in the presence of copper. These plasmids (pFBA20) were tentatively classed as cryptic (80). Without any apparent selection, the plasmids were maintained within the stream system for more than 1 year. Identical plasmids were found in morphologically and phenotypically indistinguishable hosts suggesting that the population was adapted to long-term survival and was capable of being dispersed. In addition, identical plasmids were found in two phenotypically distinguishable hosts indicating that the plasmids had at some point transferred. This presents contrary evidence to the hypothesis that under no apparent selection cryptic plasmids should be lost due to the metabolic burden they place on the cell in a potentially hostile environment (11, 90, 94).

A survey of antibiotic resistant bacteria isolated from freshwater provides a cautionary note to extrapolating gene transfer events from phenotypic analysis of multi-antibiotic resistance (48, 49). There are numerous reports where plasmid detection is linked with particular phenotypic characters and the implication is that the plasmids encode these characters (for example, Refs 26, 40). In a study of antibiotic resistance the incidences of single and multiple resistance were compared between sewage outfall into Windermere (Cumbria, UK), the lake itself and two remote upland tarns which were not influenced by sewage treatment practices. A preliminary survey indicated that Windermere showed a higher incidence of resistance in bacteria isolated from lakewater than from those isolated from the sewage effluent, discharging into the lake. More surprising was the finding that the incidence of bacterial resistance was even higher in the two remote upland tarns. These differences could have arisen due to differences in species composition, which greatly influences the sensitivity of the test results. For example, the antibiotic resistance profiles of the pseudomonad population isolated from the lake and from the effluent were similar although the species composition differed greatly. Profiles of individual *Pseudomonas* species revealed considerable variation. Therefore the species composition must be determined before any differences can be determined sensibly. Resistance has also been associated with growth in low nutrients due to changes in cell membrane composition (15). Cells showing multiple resistances could be rendered sensitive by increasing the nutrient concentration in batch or continuous culture (50). The increase in sensitivity was accompanied by major changes in the outer membrane structure (50). The evidence presented by retrospective analysis supports the generalization that plasmids are ubiquitous, and misinterpretation of their function can be avoided by the use of *in vitro* transfer experiments (40).

IN SITU AND RETROSPECTIVE GENE TRANSFER ANALYSIS: POTENTIAL FOR THE TRANSFER OF RECOMBINANT GENES IN THE ENVIRONMENT

The available evidence gained through both *in situ* and retrospective transfer analysis suggests that genetic transfer by a trichotomy of processes can occur in a range of diverse environments involving large numbers of different bacterial hosts representing inter-species and inter-generic transfer (60, 92, 111). There is also evidence that gene transfer can cross inter-microbial boundaries, even between distantly related bacteria, for example from *E. coli* to *Saccharomyces cerevisiae* (18, 67). The background of information suggests that transfer events can occur naturally and without apparent selection although when selective pressures are applied the frequency will increase.

Much information has been obtained from the two experimental approaches described in this chapter (Table 3.3). *In situ* experimentation places its emphasis on a quantitative assessment of transfer. Its future role should be targeted towards gathering information on the potential for transfer from a released organism into the indigenous population. In contrast, retrospective analysis can only qualitatively assess transfer events which, by definition, have already occurred. There is still much to be gained from plasmid epidemiology about the evolution of plasmids as they spread through indigenous bacterial populations (see Table 3.3). There is no reason to suppose that GEMs carrying recombinant DNA will behave differently. Given that the estimated transfer frequency in the environment is low then other factors will affect greatly the dissemination of recombinant genes. It is the fate of the host bacterium that is critical in evaluating the potential fate of the introduced gene. Extensive reviews on monitoring recombinant bacteria in the environment show that many of the molecular techniques are

Table 3.3 Comparison of *in situ* and retrospective plasmid transfer experimentation.

In situ:
  Shows that plasmid transfer occurs in the aquatic environment
  Allows quantitative assessment of transfer frequency
  Defines the effects of a limited number of single parameters on transfer frequency
  Can be used with an extensive range of bacteria and plasmids
  Method of assessment required for monitoring transfer into indigenous
    populations

Retrospective:
  Implies that plasmid transfer has occurred
  Allows only qualitative assessment of transfer
  Allows deductions to be made regarding structural evolution of plasmids

---

applicable to the study of plasmid transfer in a variety of habitats (24, 44, 82, 83).

Whether we can realistically predict the transfer potential of a recombinant gene in the environment remains to be seen. Meanwhile, careful genetic manipulation of bacteria with the potential to be released can reduce the possibility of uncontrolled dissemination of their recombinant DNA in the environment. The insertion of recombinant genes onto the chromosome of the release host would preclude the use of plasmids. The rate of mobilization of chromosomal genes is several orders lower than that of the transfer of conjugative and mobilizable plasmids. If the gene has no selective advantage then its chance of becoming predominant is reduced. Therefore, linking recombinant genes to those encoding antibiotic resistance is inadvisable. The development of suicide systems where an organism 'self-destructed' after completing its task would further reduce the likelihood of DNA transfer (10, 17). Bej *et al.* (10) constructed a suicide plasmid comprising the *hok* gene, which encodes a lethal polypeptide, under the control of the *lac* promoter. As with systems in which the cell contains a constitutively expressed lethal gene protected by an externally induced protective gene (which switches off in the absence of the inducer) (17), both can suffer from spontaneous mutation which renders the suicide system ineffective. After release, a sub-population of cells was detected which had become resistant to the polypeptide of the *hok* system. Despite these initial difficulties, the use of bacterial suicide systems may become a useful method for biocontrol of GEMs.

In a report to the Ecological Society of America, Tiedje and colleagues (109) concluded that:

the available scientific evidence indicates that lateral transfer among micro-organisms in nature is neither so rare that we can ignore its occurrence, nor so common that we can assume that the barriers crossed by biotechnology are comparable to those constantly crossed in nature

Multidisciplinary research combining the modern techniques of molecular biology and bacterial genetics combined with the traditional, but essential, discipline of microbial ecology may provide a satisfactory solution to this dilemma.

ACKNOWLEDGEMENTS

The authors would like to thank The Institute of Freshwater Ecology, the Natural Environment Research Council and The Freshwater Biological Association for support. Trevor Furnass is thanked for the photographs.

REFERENCES

1. Abelson, J. (1977) *Science* **196**, 159–160.
2. Alexander, M. (1985) In *Potential Impacts of Environmental Release of Biotechnology Products: Assessment, Regulation and Research Needs*, Eds Gillett, J.W., Stern, A.M., Levin, S.A., Harwell, M.A., Alexander, M. & Andow, D.A. Cornell University, New York, pp. 63–77.
3. Altherr, M.R. & Kasweck, K.L. (1982) *Applied and Environmental Microbiology* **44**, 838–843.
4. Amin, M.K. & Day, M.J. (1988) *Letters in Applied Microbiology* **6**, 93–96.
5. Babich, H. & Stotzky G. (1980) *Water Research* **14**, 185–187.
6. Bale, M.J., Fry, J.C. & Day, M.J. (1987) *Journal of General Microbiology* **133**, 3099–3107.
7. Bale, M.J., Fry, J.C. & Day, M.J. (1988) *Applied and Environmental Microbiology* **54**, 972–978.
8. Barkay, T. (1987) *Applied and Environmental Microbiology* **53**, 2725–2742.
9. Barnthouse, L.W. & Palumbo, A.V. (1986). In *Biotechnology Risk Assessment 1985*, Eds Fiskel, J. & Covello, V.T. Pergamon Press, Oxford, pp. 109–128.
10. Bej, A.K., Perlin, M.H. & Atlas, R.M. (1988) *Applied and Environmental Microbiology* **54**, 2472–2477.
11. Bennet, P.M. & Linton, A.G. (1986) *Journal of Antimicrobial Chemotherapy* **18** (Suppl. C), 123–126.
12. Bergh, O., Borsheim, K.Y., Bratbak, G. & Heldal, M. (1989) *Nature* **340**, 467–468.
13. Bloom B. R. (1980) *Nature* **284**, 593–595.
14. Brill, W.J. (1988) *Issues in Science & Technology* **4**(3) 44–50.
15. Brown, M.R.W. (1977) *Journal of Antimicrobial Chemotherapy* **3**, 198–201.
16. Burton, N., Day, M.J. & Bull, A.T. (1982) *Applied and Environmental Microbiology* **44**, 1026–1029.
17. Cuskey, S.M. (1988) Roundtable 3: Survival, persistence and colonization. In *The Release of Genetically Engineered Microorganisms*, Eds Sussman, M., Collins, C.H., Skinner, F.A. & Stewart-Tull, D.E. Academic Press, London, pp. 233–234.
18. Davies, J. (1990) *Trends in Biotechnology*, **8**, 198–203.
19. Day, M.J. (1987) *Science Progress* **71**, 203–220.

20. DeFlaun, M.F., Paul, J.H. & Davis, D. (1986) *Applied and Environmental Microbiology* **52**, 654–659.
21. DeFlaun, M.F., Paul, J.H. & Jeffrey, W.H. (1987) *Marine Ecology Progress Series* **38**, 65–73.
22. Dorward, D.W. & Garon, C.F. (1989) *Journal of Bacteriology* **171**, 4196–4201.
23. Duboise, S.M., Moore, B.E., Sorber, C.A. & Sagik, B.P. (1979) *Critical Reviews of Microbiology* **7**, 245–285.
24. Ford, S.F. & Olsen, B. (1988) Methods for detecting genetically engineered microorganisms in the environment. *Advances in Microbial Ecology* **10**, 45–79.
25. Fernandez-Astorga, A., Meula, A., Cisterna, R., Iriberri, J. & Barcini, I. (1992) *Advances in Microbial Ecology* **10**, 45–57.
26. Fredrickson, J.K., Hicks, R.J., Li, S.W. & Brockman, F.J. (1988) *Applied and Environmental Microbiology* **54**, 2916-2923.
27. Fry, J.C. & Day, M.J. (Eds) (1990) *Bacterial Genetics in Natural Environments*. Chapman & Hall, London.
28. Fry, J.C. & Day, M.J. (1990) In *Bacterial Genetics in Natural Environments*, Eds Fry, J.C. & Day, M.J. Chapman & Hall, London, pp. 55–80.
29. Gealt, M.A., Chai, M.D., Alpert, K.B. & Boyer, J.C. (1985) *Applied and Environmental Microbiology* **49**, 836–841.
30. Genther, F.J., Chatterjee, P., Barkay, T. & Bourquin, A.W. (1988) *Applied and Environmental Microbiology* **54**, 115–117.
31. Glassman, D.L. & McNicol, L.A. (1981) *Plasmid* **5**, 231–236.
32. Ghosal, D., You, I.S., Chatterjee, O.K. & Chakrabarty, A.M. (1985) *Science* **228**, 135–142.
33. Gowland, D.C. & Slater, J.H. (1984) *Microbial Ecology* **10**, 1–13.
34. Grabow, W.O.K., Prozesky, O.W. & Burger, J.S. (1975) *Water Research* **9**, 777–782.
35. Guerry, P. & Colwell, R.R (1977) *Infection & Immunity* **16**, 328–334.
36. Hada, H.S. & Sizemore, R.K. (1981) *Applied and Environmental Microbiology* **41**, 199–202.
37. Hall G.H., Jones, J.G., Pickup, R.W. & Simon, B.M. (1990) *Methods in Microbiology* **23**, 181–210.
38. Hansen, J. B. & Olsen, R.H. (1978) *Journal of Bacteriology* **135**, 227–238.
39. Hansen, C.L., Zwolinsk, G., Martin, D. & Williams, J.W. (1984) *Biotechnology and Bioengineering* **26**, 1330–1333.
40. Hermansson, M., Jones, G.W. & Kjelleberg, S. (1987) *Applied and Environmental Microbiology* **53**, 2338–2342.
41. Hughes, C. & Meynell, G.G. (1974) *Lancet* **ii**, 451–453.
42. Iriberri, J.M., Unanue, M., Barcina, I. & Egea, L. (1987) *Applied Environmental Microbiology* **53**, 2308–2314.
43. Jain, R.K., Sayler, G.S., Wilson, J.T., Houston, L. & Pacia, D. (1987) *Applied and Environmental Microbiology* **53**, 996–1002.
44. Jain, R.K., Burlage, R.S. & Sayler G.S. (1988) *Critical Reviews in Biotechnology* **8**, 33–84.
45. Jobling, M.G., Peters, S.E. & Ritchie, D.A. (1988) *FEMS Microbial Ecology* **49**, 31–37.
46. Jones, J.G. (1977) *Freshwater Biology* **7**, 67–91.
47. Jones, J.G., Orlandi, M.J.L.G. & Simon, B.M. (1979) *Journal of General Microbiology* **115**, 37–48.

48. Jones, J.G., Gardener, S., Simon, B.M. & Pickup, R.W. (1986) *Journal of Applied Bacteriology* **60**, 443–453.
49. Jones, J.G., Gardener, S., Simon, B.M. & Pickup, R.W. (1986) *Journal of Applied Bacteriology* **60**, 455–462.
50. Jones, J.G. & Pickup, R.W. (1989) *Aqua* **38**, 131–135.
51. Kahn, M.E., Barany, F. & Smith, H.O. (1983) *Proceedings of the National Academy of Sciences of the USA*, **80**, 6927–6931.
52 Keeler, K.H. 1989. Critical Review in Biotechnology **8**, 85–97.
53. Kelly, W.J. & Reanney, D.C. (1984) *Soil Biology & Biochemistry* **16**, 108.
54. Keil, H., Keil, S., Pickup, R.W., & Williams, P.A. (1985) *Journal of Bacteriology* **164**, 887–895.
55. Khalil, T.A. & Gealt, M.A. (1987) *Canadian Journal of Microbiology* **33**, 733–737.
56. Khesis, R.B. & Karasyova, E.C. (1984) *Molecular and General Genetics* **197**, 280–285.
57. Kobori, H.C., Sullivan, C.W. & Shizuya, H. (1984) *Applied and Environmental Microbiology* **48**, 515–518.
58. Levin, B.R., Stewart, F.M. & Rice, V.A. (1979) *Plasmid* **2**, 247–260.
59. Levin, S.A. & Harwell, M.A. (1985) In *Potential Impacts of Environmental Release of Biotechnology Products: Assessment, Regulation and Research Needs*, Eds Gillett, J.W., Stern, A.M., Levin, S.A., Harwell, M.A., Alexander, M. & Andow, D.A. Cornell University, New York, pp. 134–173.
60. Levy, S.B. & Marshall, B.M. (1988) In *The Release of Genetically Engineered Micro-organisms*, Eds Sussman, M., Collins, C.H., Skinner, F.A., & Stewart-Tull, D.E. Academic Press, London, pp. 61–76.
61. Levy, S.B. & Miller, R.V. (1989) *Gene Transfer in the Environment*. McGraw-Hill, New York.
62. Lock, M.A., Wallace, R.R., Costerton, J.W., Ventullo, R.M. & Charlton, S.E. (1984) *Oikos* **42**, 10–22.
63. Lynch, J.D. & Hobbie, J.E. (1988) *Microorganisms in Action: Concepts and Applications in Microbial Ecology*. Blackwell Scientific Publications, Oxford.
64. Mach, P.A. & Grimes, D.J. (1982) *Applied & Environmental Microbiology* **44**, 1395–1403.
65. Mancini, P., Fertris, S., Nave, D. & Gealt, M.A. (1987) *Applied & Environmental Microbiology* **53**, 665–671.
66. Maniatis, T., Fritsch, E.F. & Sanbrook, J. (1982) *Molecular Cloning*. Cold Spring Harbor, New York.
67. Mazodier, P. & Davies, J. (1991) *Annual Review of Genetics* **25**, 147–151.
68. McClure, N.C., Weightman, A.J. & Fry, J.C. (1989) *Applied and Environmental Microbiology* **55**, 2627–2634.
69. McClure, N.C., Fry, J.C. & Weightman, A.J. (1989) *Applied and Environmental Microbiology* **57**, 366–373.
70. McFeters, G.A. & Stuart, D.G. (1972) *Applied and Environmental Microbiology* **24**, 805–811.
71. Miller, R.V. & Levy, S.B. (1989) In *Gene Transfer in the Environment*, Eds Levy, S.B. & Miller, R.V. McGraw-Hill, New York.
72. Morgan, J.A.W., Winstanley, C., Pickup, R.W., Jones, J.G. & Saunders, J.R. (1989) *Applied and Environmental Microbiology* **55**, 2537–2544.

73. Neuhold, J. & Ruggiero, L. (1975) National Science Foundation Report (NSF-RA-76008) Washington, 44 pp.
74. O'Morchoe, S.B., Ogaunseitan, O., Sayler, G.S. & Miller, R.V. (1988) *Applied and Environmental Microbiology* **54**, 1923–1929.
75. Paul, J.H. & David, A.W. (1989) *Applied and Environmental Microbiology* **55**, 1865–1869.
76. Paul, J.H., DeFlaun, M.F., Jeffrey, W.H. & David, A.W. (1988) *Applied and Environmental Microbiology* **54**, 718–827.
77. Paul, J.H., Jeffrey, W.H., David, A.W., DeFlaun, M.F. & Cazares, L.H. (1989) *Applied and Environmental Microbiology* **55**, 1823–1828.
78. Paul, J.H., Jeffrey, W.H. & DeFlaun, M.F. (1987) *Applied and Environmental Microbiology* **53**, 170–179.
79. Pickup, R.W. & Williams, P.A. (1982) *Journal of General Microbiology* **128**, 1385–1390.
80. Pickup, R.W. (1989) *Microbial Ecology* **18**, 211–220.
81. Pickup, R.W., Saunders, J.R., Morgan, J.A.W., Winstanley, C., Jones, J.G., Simon, B.M., *et al.* (1989) In *Watershed 1989*, Eds. Wheeler, D., Richardson, & Bridges, J. Pergamon Press, Oxford, pp. 375–380.
82. Pickup, R.W. & Saunders, J.R. (1990) *Trends in Biotechnology* **8**, 329–334.
83. Pickup, R.W. (1991) *Journal of General Microbiology* **137**, 1009–1019.
84. Poole, N.J. (1988) In *The Release of Genetically Engineered Microorganisms*, Eds. Sussman, M.J., Collins, E.H., Skinner, F.A. & Stewart-Tull, D.E. Academic Press, London, pp. 265–274.
85. Powledge, T.M. (1983) *Biotechnology* **9**, 743–755.
86. Proctor, L.M. & Fuhrman, J.A. (1990) *Nature* **342**, 60–61.
87. Reanney, D.C., Gowland, P.C., & Slater, J.H. (1983) *Genetic Interactions among Microbial Communities. Symposium of Society of General Microbiology.* Vol. 34, pp. 379–421. IRL Press, Oxford.
88. Rochelle, P.A., Fry, J.C. & Day, M.J. (1989) *Journal of General Microbiology* **135**, 409–424.
89. Roszak, D.B. & Colwell, R.R. (1987) *Microbiological Reviews* **51**, 365–379.
90. Saunders, J.R. (1984) *British Medical Bulletin* **40**, 54–60.
91. Saye, D.J. & Miller, R.V. (1989) In *Gene Transfer in the Environment* Eds Levy, S.B. & Miller, R.V. McGraw-Hill, New York, pp. 223–260.
92. Saye, D.J., Ogenseitan, O., Sayler, G.S. & Miller, R.V. (1987) *Applied and Environmental Microbiology* **53**, 987–995.
93. Schilf, W. & Klingmuller, W. (1983) *Recombinant DNA Technology Bulletin* **6**, 101–102.
94. Simonsen, L. (1991) *Microbial Ecology* **22**, 187–205.
95. Schutt, C. (1989) *Microbial Ecology* **17**, 49–62.
96. Simberloff, D. (1981) In *Biotic Crises in Ecological and Evolutionary Time*, Ed. Nitecki, M.H. Academic Press, New York, pp. 53–81.
97. Simon, R.D., Shilo, W. & Hastings, J.W. (1982) *Current Microbiology* **7**, 175–180.
98. Sinclair, J.L. & Alexander, M. (1989) *Canadian Journal of Microbiology* **35**, 578–582.
99. Singleton, F.L. & Guthrie, R.K. (1977) *Water Research* **11**, 639–642.
100. Smith, H. (1989) *Annual Review of Microbiology* **43**, 1–22.

101. Smith, H.W. (1970) *Nature* **228**, 1296–1298.
102. Stewart, G.J. & Carlson, C.A. (1986) *Annual Review of Microbiology* **40**, 211–237.
103. Stotzky, G. & Babich, H. (1984) *Recombinant DNA Technology Bulletin* **7**, 163–188.
104. Stotzky, G. & Babich, H. (1986) *Advances in Applied Microbiology*, **31**, 93–138.
105. Sturtevant, A.B.Jr & Feary, J.W. (1969) *Applied and Environmental Microbiology* **18**, 918–924.
106. Sussman, M., Collins, C.H., Skinner, F.A. & Stewart–Tull, D.A. (Eds) (1988) *The Release of Genetically Engineered Micro-organisms.* Academic Press, London.
107. Taub, F.E. (1976) *International Journal of Environmental Studies* **10**, 23–33.
108. Tauxe, R.V., Holmberg, S.D. & Cohen, M.L. (1989) In *Gene Transfer in the Environment* Eds, Levy, S.B. & Miller, R.V. McGraw-Hill, New York, pp. 377–404.
109. Tiedje, J.M., Colwell, R.K., Grossman, Y.L., Hodson, R.E., Lenski, R.E., Mack, R.N. *et al.* (1989) *Ecology* **70**, 298–315.
110. Trevors, J.T. (1985) *Microbiological Sciences* **3**, 259–271.
111. Trevors, J.T., Barkay, T. & Bourquin, A.W. (1987) *Canadian Journal of Microbiology* **33**, 191–198.
112. Trevors, J. T. (1988) *Microbiological Sciences* **5**, 132–136.
113. Trevors, J.T. & Elvan Elsas, J.D. (1989) *Canadian Journal of Microbiology* **35**, 895–902.
114. Walter, M.V., Porteous, A. & Seidler, R.J. (1987) *Applied and Environmental Microbiology* **53**, 105–109.
115. Wheatcroft, R.W. & Williams, P.A. (1981) *Journal of General Microbiology* **124**, 433–437.
116. Wickham, G.S. & Atlas, R.M. (1988) *Applied and Environmental Microbiology* **54**, 2192–2196.
117. Willetts, N. & Wilkins, B. (1984) *Microbiological Reviews* **48**, 24–41.
118. Woodruff, H.B. (1980) *Science* **208**, 1225–1229.
119. Wortman, A.T. & Colwell, R.R. (1988) *Applied and Environmental Microbiology* **54**, 1284–1288.

Chapter 4

Composting as a Model System for Monitoring the Fate of Genetically Manipulated Gram-positive Bacteria

W. Amner, C. Edwards and A.J. McCarthy

Department of Genetics & Microbiology, University of Liverpool

INTRODUCTION

The proposed release and potential establishment of genetically engineered microorganisms (GEMs) in the environment has given rise to particularly intense debate about the possible risks involved. Among those risks frequently discussed are the displacement of indigenous microbial populations, disruption of ecological cycles and undesired transfer of novel genetic traits to other species (32, 88). These concerns have prompted the development of laboratory-contained microcosms as model environments for risk-assessment studies. While such model systems cannot allow accurate prediction of the

Monitoring Genetically Manipulated Microorganisms in the Environment. Edited by C. Edwards
Published 1993 John Wiley & Sons Ltd. © 1993 W. Amner, C. Edwards and A.J. McCarthy

fate of GEMs in the environment, they can be used to generate considerable data for modelling without the regulatory constraints of field testing.

A variety of microcosms have been developed and employed for assessing the survival of and potential for gene transfer between introduced bacteria, and some of these are listed in Table 4.1. They provide controlled systems for studying the effects of environmental variables on the fate of both microorganisms and DNA in the environment (14, 87, 104, 106, 107), and for studying the efficacy of GEMs under environmental conditions (41, 78). Whilst the choice of microcosm and introduced organism(s) have varied, the majority of studies have focused on survival and gene exchange among Gram-negative species, particularly the pseudomonads, with an emphasis on conjugal gene transfer. In these studies antibiotic resistances have generally been used as selective markers for introduced donors and potential recipient and transconjugant cells. Gene transfer to species indigenous to microcosms has been more difficult to assess due to the problem of donor counter selection, and is infrequently investigated by researchers (2, 25, 108). As a result, reports of gene transfer to indigenous populations appear to be limited to mobilization of non-conjugative plasmids to indigenous recipients in an aquatic microcosm (37, 71). Another area of research which has received little attention concerns the effect of GEMs on indigenous populations and ecological cycles. The effect of recombinant polygalacturonide-producing *Pseudomonas* spp. on wild-type *Pseudomonas* growth was recently studied in rhizosphere and non-rhizosphere soils (118) and the potential for *Azospirillum* Tn5 mutants to displace rhizosphere populations of nitrifying bacteria was investigated in intact soil microcosms (11). In addition the effect of genetically engineered *Erwinia caratovora* on functional groups of indigenous bacteria was studied in an aquatic microcosm (93). In all of these studies no significant ecological effects were observed. Reports of measurable effects of GEMs on ecological processes are currently limited to the increase in carbon mineralization by a recombinant *Streptomyces lividans* strain released into soil (111).

The success of studies aimed at investigating the effects of introduced GEMs on indigenous populations and ecological cycles is dependent on the ability to identify the species within microbial populations and monitor different microbial processes such as nitrification. Detection of introduced organisms may be aided by the use of cloned marker genes such as antibiotic resistance genes, fluorescent *lacZY* genes (26), bioluminescence genes (95), red pigmentation (prodigiosin) genes (21), the *xylE* gene that encodes catechol 2,3 dioxygenase (116) and the *pglA* gene encoding polygalacturonidase (118). Detection of indigenous organisms, on the other hand, has generally been dependent on selective cultural techniques and on fluorescent antibody and other direct microscopic methodologies. More recently DNA probes, enzyme assays and methods for the detection of specific antibodies have been used to detect specific bacterial populations in environmental samples (10, 12, 50, 59). However, such techniques are highly specific and limited to a narrow range of indigenous species. At present cultural techniques appear to be the most

Table 4.1 Examples of microcosms used for risk assessment studies.

Microcosm	Introduced organism(s)	Reference
Soil slurry	Pseudomonas cepacia	110
Soil sub-sample	Escherichia coli	60
	Bacillus cereus, Bacillus subtilis	117
	E. coli, Rhizobium fredii	87
	Pseudomonas spp.	104
	Streptomyces spp.	85, 86, 111, 113
	E. coli	39, 119
	B. subtilis	44
Intact soil core	Pseudomonas sp. (non-fluorescent)	35
	Azospirillum lipoferum	11
Soil system	Actinophage, streptomycetes	19
Rhizosphere	Pseudomonas sp. (fluorescent)	108, 109
	Pseudomonas aeruginosa, Pseudomonas putida	118
	E. coli, P. putida	80
Mushroom compost	Saccharomonospora viridis	2
	B. subtilis	4
Activated sludge unit	P. putida	78
Agricultural drainage water	P. fluorescens, P. putida, Klebsiella aerogenes	106
Lakewater test chambers	P. aeruginosa	92
River stones placed in epilithon	P. aeruginosa	9
Inoculated slate disc in rotating disc chemostat	Pseudomonas spp.	89
Red clover	Rhizobium leguminosarum bv. trifoli	72
Water and sediment	Natural population	48
	Pseudomonas spp.	112
Lake and sewage water samples	E. coli	18
Marine water	E. coli, Salmonella enteritidis	86

sensitive and widely applicable of detection methods. One environment where these techniques have been studied in detail for a number of bacterial groups is mushroom compost (3). This substrate has been selected and developed as a model environment for GEM risk assessment studies and will be discussed at length in this chapter. The properties of this natural substrate will be discussed in relation to predicting the consequences of the release of GEMs into the environment. Emphasis will be placed on the study of thermophilic Gram-positive bacteria, which are abundant in mushroom compost. Thermophiles are here defined as those organisms showing substantial growth at or above 50°C and therefore include many organisms (particularly the actinomycetes) which tend to be thermotolerant.

COMPOSTING

Composting is an aerobic process which involves the thermophilic treatment of solid-state organic materials by microbial activities. The material being composted provides its own source of nutrients, moisture and microbial inoculum. A high temperature is generated by self-heating. Composting practices range from the traditional treatment of agricultural wastes to provide composts for soil amendment or substrates for mushroom cultivation to the more recent treatment of municipal solid waste and wastewater (sewage) sludge to render them environmentally acceptable (33). Mushroom compost provides a unique model for risk assessment purposes by providing a matrix which is environmentally complex but has a biological and microbial composition controlled by the compost preparation process. The major advantages of using this substrate as a model microcosm system are as follows:

1. Mushroom compost contains a *diverse* and highly *active* population of mainly Gram-positive bacteria.
2. It provides an abundant source of many industrially important bacteria including thermophilic actinomycetes and *Bacillus* spp. These organisms have received little attention in risk assessment studies in spite of the potential for strain improvement and increased industrial and envrionmental application through recombinant DNA technology.
3. Mushroom compost is a man-made substrate and is therefore readily available in a relatively reproducible form compared to natural solid-state substrates and is therefore amenable to both containment and scale-up.
4. Mixing of the compost during preparation prevents the development of severe environmental gradients and imparts a degree of homology to an otherwise heterogeneous substrate.
5. Exposure of the compost to high temperatures during the pasteurization phase results in a relatively predictable biological community.
6. Since composting is initiated by growth of mesophilic microorganisms, mushroom compost contains many mesophilic strains and provides a

substrate for studying potential gene transfer between mesophilic and thermophilic populations.

Commercial preparation of mushroom compost

The commercial preparation of mushroom compost is carried out following traditional methods based on a two-phase fermentation of stable bedding (mixtures of wheat straw and horse manure) supplemented with chicken manure (30). During the initial phase of preparation, Phase 1, the pre-wetted raw materials are stacked into long piles (approximately 1.5 m high) and left to self-heat outdoors for 7–10 days. Spontaneous heat generation is sufficient to achieve temperatures of around 70°C in the centre of the piles. The compost stack is turned at intervals during this phase to prevent the development of anaerobic conditions. This process also helps to minimize the establishment of severe environmental gradients and contributes a degree of homology to the substrate. In Phase II the compost is spread onto trays in insulated rooms, rapidly heated to 60°C then maintained at around 50°C for a further 7 days. This second phase of composting, the pasteurization or peak heat phase, is designed to drive off excess ammonia from the substrate and to eliminate potential pathogens and competitors of the subsequent mushroom crop. The compost is then ready to be spawned with the edible mushroom *Agaricus bisporus*.

Mushroom compost collected at the end of Phase II of composting provides a relatively reproducible model substrate for risk assessment studies. The substrate is homogenized to an extent by mechanical mixing in Phase I and the biological community is controlled by the pasteurization phase. Biological activity is also expected to be high due to the addition of carbon and nitrogen sources in the form of molasses, prior to Phase I of preparation.

Microbiology of mushroom compost

Mushroom compost provides an abundant source of many industrially important groups of microorganisms, notably the thermophilic actinomycetes and *Bacillus* spp. The actinomycete population has been characterized in a number of studies due to the presence of many species of lignocellulolytic importance and due to the role of some species in hypersensitivity pneumonitis (63, 66, 75). The occurrence and predominance of different microbial groups change during preparation giving rise to a relatively predictable succession of microorganisms (30, 64, 99). Phase I is characterized by a transition from mesophilic microorganisms to thermophilic fungi, e.g. *Humicola* spp. and *Chaetomium thermophile*, and to spore-forming bacteria, predominantly *Bacillus* spp. and thermophilic actinomycetes belonging to the genera *Thermomonospora*, *Thermoactinomyces* and *Streptomyces*, which become increasingly dominant during the pasteurization phase producing a

Table 4.2 Microorganisms frequently isolated at or above 50°C from mushroom compost.

Fungi	Actinomycetes	Other bacteria
Humicola grisea	Thermomonospora spp. (white)	Bacillus subtilis
Chaetomium thermophile	Thermomonospora chromogena	B. coagulans
Aspergillus fumigatus		
	Thermoactinomyces vulgaris	B. licheniformis
	Streptomyces spp. (grey)	Bacillus stearothermophilus
	Saccharomonospora viridis	Pseudomonas sp
	Feania rectivirgula	

characteristic whitish-grey film or 'fire-fang' over the substrate. The diversity of thermotolerant and thermophilic microorganisms frequently isolated from mushroom compost is shown in Table 4.2. The numbers of these organisms are high with end of Phase II samples yielding up to 10^{10} thermophilic bacilli and 10^8 thermophilic actinomycetes per gram (3). These groups include both spore-formers and endospore-formers and therefore have a great capacity for survival in the environment as dormant forms. The potential environmental impact of such organisms following receipt of genetic material from released GEMs is thus increased due to their potential longevity as spores. Although generally considered as an environment where oxygen is readily available, Derikx et al. (22) recently detected methane in air emitted from Phase I compost piles and isolated high numbers of the methanogenic bacterium Methanobacterium thermoautotrophicum. The populations isolated were approximately equal in number to the aerobic thermophilic population and were thought to be associated with facultatively anaerobic bacteria providing protection against oxygen. The presence of methanogens in compost increases the diversity of the total microbial community, widens the scope for gene survival and transfer studies and indicates the potential for the isolation of novel species.

ISOLATION OF THERMOPHILIC BACTERIA FROM COMPOST

Cultural techniques: thermophilic actinomycetes

Use of plating techniques to isolate microorganisms from environmental samples has long been regarded as a problem because of the presence of non-culturable cells (34, 90, 105). However, much research has been directed at improving the recovery of thermophilic actinomycetes from overheated substrates using conventional plating methods (74). Colony-forming ability therefore remains a relatively sensitive method for isolating actinomycetes from compost.

Isolation media typically contain cycloheximide to inhibit fungal growth and are maintained at a high pH to suit the neutral to alkaline pH requirement of most actinomycetes. Incubation temperatures are limited to 50–60°C since most actinomycete species are regarded as only moderately thermophilic.

Dilution plating

Recovery of thermophilic actinomycetes using conventional dilution plating techniques is hindered by the tendency for faster growing *Bacillus* spp. to submerge and inhibit the development of actinomycete colonies. As a result enumeration and identification of actinomycete groups is severely restricted. The employment of alternative diluents and extraction procedures appears to be ineffective in overcoming this problem (3). Figure 4.1a shows a dilution

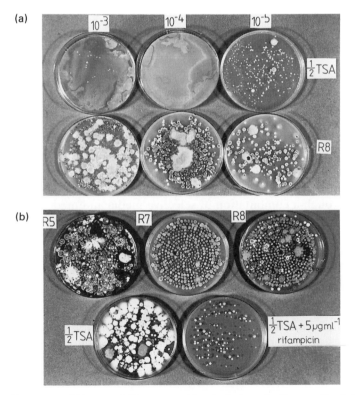

Fig. 4.1 Recovery of thermophilic compost bacteria by selective media. (a) Compost extract was serially diluted and 10^{-3}, 10^{-4}, 10^{-5} dilutions plated onto either half-strength tryptone soya agar (½ TSA) or R8 medium. (b) Improved and selective recovery of thermophilic actinomycetes using an Andersen sampler stack. The same compost sample was used to isolate thermophilic bacteria. The data illustrate diversity of actinomycete species that can be revealed by using different selective media.

Table 4.3 Selective isoloation of thermophilic actinomycete groups from mushroom compost.

Selective medium or supplement	Actinomycete group(s) selected	Reference
R8 medium	*Saccharomonospora viridis, Faenia rectivirgula*	2
Proline-salts medium	*Streptomyces thermoviolaceus*	Featherstone & Edwards (unpublished work)
Casein hydrolysate (0.2% w/v)	*F. rectivirgula*	67
Novobiocin (25 μg ml^{-1})	*Thermoactinomyces*	20
Kanamycin (25 μg ml^{-1})	*Thermomonospora chromogena*	76
Rifampicin (5 μg ml^{-1})	*T. chromogena*, white *Streptomyces*	2, 6
Sodium chloride	*F. rectivirgula*	75
Hippurate	*F. rectivirgula*	43

series from a compost extract plated onto half-strength tryptone soya agar. Thermophilic bacilli predominate and the 10^{-3} and 10^{-4} dilutions illustrate the tendency of the bacilli to swarm across the plate to produce mucoid confluent growth that prevents the growth and isolation of actinomycetes. A range of selective media has been developed for increasing the specific recovery of different actinomycete groups from the environment (115). Several of these are applicable to the recovery of thermophilic actinomycetes found in mushroom compost (Table 4.3). Specific recovery of different taxa is achieved through a combination of selective media and media amended with selective inhibitors, particularly antibiotics, to remove or reduce unwanted competitors. In some cases (notably the use of R8 medium and novobiocin amended media) the numbers of bacilli and other unwanted bacteria are reduced sufficiently to enable the dilution plate count technique to be used to enumerate specific actinomycete groups (2). Figure 4.1b shows the way in which R8 reduces the problem of overgrowth by bacilli allowing isolation of actinomycetes using dilution plating.

Cocktails of antibiotics may also be added to isolation media to select for introduced strains carrying marker genes expressing multiple antibiotic resistance. This technique has been successfully employed in the detection of recombinant mesophilic *Streptomyces* strains released into soil (14, 85, 111, 113).

Sedimentation chamber/Andersen sampler methods

A more efficient recovery regime for actinomycetes makes use of an Andersen sampler in combination with a wind tunnel (45), or sedimentation chamber (68). In the latter method dried compost samples are prepared, which serves to kill off most vegetative cells. The samples are then agitated within a sedimentation chamber to produce a cloud of spores. *Bacillus* spores tend to clump together or adhere to the dried substrate and are therefore not maintained in suspension. Air samples drawn from the chamber after a period of sedimentation are rich in actinomycete spores, which may be recovered by passage through an Andersen sampler stacked with appropriate isolation plates. Each plate in the sampler stack is separated by a metal disc; these discs have graduated hole sizes such that the disc bearing the smallest hole is at the bottom of the stack. Actinomycete colonies are concentrated on plates from the lowest stages of the sampler (stages 5 and 6) where the critical velocity for impaction of actinomycete spores is attained. Figure 4.1b shows the applicability of this method. Plates of different isolation media taken from the lowest stages of the stack illustrate (i) how actinomycetes predominate and (ii) the different types of actinomycetes that may be isolated by choosing an appropriate growth medium.

Using this method thermophilic actinomycetes are recovered as discrete colonies. Many can be tentatively identified to the generic level based on distinctive morphological characteristics using a microscope fitted with a high-power long working objective. Accurate identification requires the use of numerical phenetic, chemotaxonomic and phylogenetic data as reviewed elsewhere (42, 46, 77). The numbers of bacteria recovered are generally expressed as colony forming units (cfu) per Andersen sampler stack and are not directly comparable with results obtained by dilution plating. Statistical corrections may be necessary to account for multiple deposition of colonies at one site on an isolation plate (5). The method does, however, enable the recovery of a diverse number of actinomycete taxa and has been successfully used to monitor population changes in incubated model compost microcosms (see Fig. 4.2). Where sophisticated air sampling equipment is not available the gravity settling plate method may be used as an inexpensive and rapid alternative (103). Dried samples are simply shaken in plastic bags then puffed out into isolation plates. These approaches could be readily adapted to monitor air dispersal of GEMs.

Cultural techniques: thermophilic bacilli

Thermophilic bacilli are readily recovered in high numbers from mushroom compost onto non-selective media. Identification of different groups is complicated by the confluent colony form of many strains and the lack of morphological characteristics to aid preliminary analysis. Consequently reports aimed at identifying thermophilic bacteria, other than actinomycetes, in overheated substrates are lacking. A notable exception is the study of thermophilic microorganisms in solid waste composting by Strom (101) where major emphasis was placed on the genus *Bacillus*.

Use of cloned antibiotic resistance marker genes has been employed in the detection of recombinant mesophilic *Bacillus* strains released in both rhizosphere and non-rhizosphere soils (44, 107, 108), and expression of plasmid-encoded tetracycline resistance genes has been used to detect *Bacillus subtilis* introduced into mushroom compost (4). Marker systems are now being developed in the authors' Department for recombinant thermophilic species.

Direct detection methods

Whilst cultural methods can be used to detect a wide range of indigenous microorganisms in compost, they are limited to the isolation of viable organisms and give little indication of growth of those organisms in the environment. Studies on *Streptomyces* in soil, for example, indicate that mycelial growth is sporadic and that these organisms exist mainly as spores, probably as a result of nutrient depletion (73). Cresswell *et al.* (19) concluded that a well-developed mycelium was required for conjugation and phage infection in *Streptomyces* resident in soil. Direct microscopic observation can be used to distinguish actinomycete spores from actively growing mycelium in soils and to distinguish both structures from fungi and bacteria (7). Actinomycete growth is also visually observed in mushroom compost by the production of a whitish film known as fire-fang over the substrate. Similarly colonization and growth of a number of thermophilic actinomycete strains can be demonstrated in sterilized compost by observing the development of a mycelium on the compost following introduction of a spore inoculum. Although such observations provide an indication of active growth, they are of limited value for differentiating between different indigenous actinomycete groups, or between indigenous and introduced species.

The direct detection of Gram-positive thermophiles is expected to be improved using methods currently being developed for mesophilic strains. These include the detection of polyclonal and monoclonal antibodies specific to both indigenous and engineered species by ELISA (113), and the production of DNA and RNA probes (33, 61, 82). Hybridization to labelled probes is particularly promising but may be limited by sensitivity of the probe. In the bacilli, for instance, both cryptic and tetracycline-resistant plasmids are closely related (24, 53, 84).

COMPOST AS A MODEL MICROCOSM

For risk assessment studies aimed at determining the impact of GEMs on indigenous populations and ecological cycles, it is first necessary to monitor biological activity in the environment under study under contained laboratory conditions. Information should be obtained on both the occurrence of biological populations within the model system and on the rate of intrinsic biological processes such as nutrient cycling. Many data are now available for mini-composting systems and are currently being generated in large-scale systems under controlled environmental conditions.

Modelling population changes in mini-composting systems

Methods for recovering thermophilic bacterial populations from mushroom compost have been rigorously evaluated and the microbiology of the compost system thoroughly characterized (3). Changes in the population of thermophilic actinomycetes in mushroom compost have also been monitored during incubation of mini-composting systems under contained conditions. Figure 4.2a shows the recovery of thermophilic actinomycetes from two different

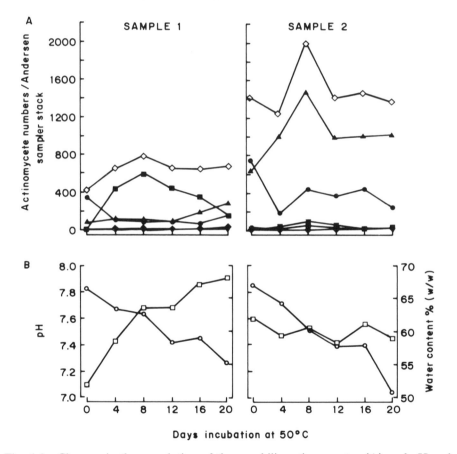

Fig. 4.2 Changes in the population of thermophilic actinomycetes (A) and pH and moisture content (B) after incubation of compost microcosms at 50°C. Total actinomycetes (◇); white *Thermomonospora* spp. (●); grey *Streptomyces* spp. (■); *Thermomonospora chromogena* (▲); *Thermoactinomyces* spp. (▼); *Saccharomonospora viridis* (◆); % (w/w) water content (○); pH (□). Values are means of triplicate samples.

batches of mushroom compost (50 g samples) incubated at 50°C for 20 days. Recoveries were made on non-selective media (1/2 strength tryptone soya agar) using the sedimentation chamber-Andersen sampler method. The initial composition of actinomycete groups in both compost samples was similar with white thermomonosporas predominating. However, during incubation of one sample the numbers of *Streptomyces* forming a grey aerial mycelium increased rapidly to become the predominant actinomycete group. In the second sample they remained a minor component of the total population and *Thermomonospora chromogena* attained dominance. These changes in population were marked and unexpected. Streptomycetes for example, are minor components of mushroom compost *per se* but are characterized by prolific spore production and rapid growth on isolation media, plus tolerance to high moisture tensions in soils (114). Members of the genus may therefore have been selectively isolated in sample 1 due to drying of the substrate during incubation (Fig. 4.2b). The lack of an apparent streptomycete population increase in the second sample is consistent with the differences in pH profiles between the two incubated systems (Fig. 4.2b). The compost samples used were collected at different times of the year, had similar initial populations and were treated under identical conditions. These results thus demonstrate the unpredictable nature of population dynamics within model systems, even when using a relatively reproducible man-made substrate.

Alternative substrates as model systems

Rapidly prepared mushroom compost

A number of experimental mushroom composts have been developed to reduce the preparation time needed to produce a suitable substrate for production of the edible mushroom (96–98). Such composts are prepared from defined mixtures of wheat straw, milk by-products, urea, peat and gypsum, and may be produced in 4–5 days compared to the usual composting period of 14–21 days. The microflora of rapidly prepared composts is similar to that of traditionally prepared composts (31), thus providing an alternative source of industrially important Gram-positive bacteria for risk assessment studies. The nutrient status of rapidly prepared composts has also been defined (98), thus providing useful background data for modelling purposes.

Alternative overheated substrates

Thermophilic actinomycetes and bacilli are ubiquitous in nature due largely to their ability to sporulate and, in the case of *Thermoactinomyces* and *Bacillus* spp., to produce endospores. They are found in particularly high numbers in a range of overheated substrates such as mouldy fodders, silage and composts (8, 27, 65, 101). Many of the thermophilic species colonizing mushroom compost are also present in these substrates. For example, *Thermoactinomyces vulgaris* and *Saccharomonospora viridis* have been isolated from hay, straw, cereal grain, animal faeces and sewage sludge compost (47, 65, 79). However, the relative abundance of the different bacterial groups varies between different substrates, with some species distinctive of, or greatly favoured by particular substrates. White thermomonosporas and *Thermomonospora chromogena* are characteristically predominant in mushroom compost, while *Streptomyces*, *Faenia rectivirgula* and *Thermoactinomyces* are prolific colonizers of mouldy hay. *Thermoactinomyces sacchari*, on the other hand, is only found in sugar cane bagasse. Differential use could be made of these substrates to provide basic model environments for studying the effect of introduced recombinant organisms on specific microbial populations.

Large scale modelling

Larger scale composting models have been described where environmental parameters can be monitored and accurately controlled. These include bench-scale continuous composting systems of 4.5 and 14 l capacity designed to simulate full-scale composting operations handling in excess of 125 tons of waste per week (49, 100).

For mushroom compost experiments a model system has been described in which compost is incubated in wooden trays (0.9 × 0.6 × 0.15 m deep) in a pasteurization chamber of 21 m³ capacity (98). The chamber has been successfully used for small-scale trials of short duration composts held under controlled environment conditions. More recently a containment facility has been developed specifically for experiments involving the release of GEMs into mushroom compost (15). Up to 40 kg of compost can be contained within a PVC-lined wire basket and held in an insulated box in a mushroom growing chamber. Environmental parameters such as temperature, aeration and moisture content can be carefully controlled and the apparatus steam sterilized and chemically disinfected at the end of each experiment.

Measuring biological activity in compost

In risk assessment studies some measurement of background biological activity is essential in order to identify disruptions to ecological processes caused by the introduction of recombinant organisms. Thus Wang *et al.* (111) were able to detect short-term increases in soil organic carbon mineralization by comparing CO_2 evolution rates from soil inoculated with a genetically engineered *Streptomyces* with those from uninoculated control soil. Several studies on composting describe techniques for monitoring a range of physical and chemical parameters which could be incorporated in release studies as indirect indicators of biological activity. These include measurements of pH (16), total nitrogen, phosphorus and gaseous ammonia (98), and organic, inorganic and volatile sulphur compounds (23). More direct measurements of biological activity could be obtained by measurements of enzyme activities (40, 102). The application of membrane inlet mass spectrometry may provide a sophisticated and sensitive biophysical technique for measuring microbial processes concomitantly and in real time. One approach is to monitor gaseous uptake or release rates as an indicator of activity, e.g. disappearance of nitrogen (N_2 fixation), NH_3 disappearance (nitrification), CH_4 evolution (methanogenesis). Such applications and their potential for studies of microbial ecology have been described by Lloyd & Scott (70).

MOLECULAR BIOLOGY OF COMPOST

Mushroom compost provides a model environment with the potential for a high frequency of gene exchange. Biological activity is high as evidenced by rapid and significant population changes during compost preparation, the development of a thick mycelial mat over the compost substrate and high enzyme activities. Dense mycelial growth ensures intimate contact between actinomycete hyphae and is likely to lead to increased DNA transfer opportunities. High biological activity is sustained by the presence of available carbon and nitrogen sources which are added to the compost in the form of molasses prior to Phase I of preparation. These nutrient amendments may also indirectly stimulate potential gene transfer events, as has been reported in other soils (14, 107).

A number of vector systems have been isolated from composts which may play a role in natural gene exchange. Pidcock *et al.* (83) screened 20 thermophilic actinomycetes from compost and isolated plasmid DNA from two thermomonospora strains. The plasmids ranged in size from 7.7 kb to greater than 25 kb and were cryptic in nature. In addition to plasmid vectors, bacteriophage-infecting strains of *Thermomonospora alba* and *T. fusca* have been isolated in abundance from compost (69). The phages showed a high degree of thermostability and appeared to be taxon specific.

Development of host marker systems

DNA screening programmes such as these provide essential information on the genetic pool within an ecosystem prior to the introduction of novel DNA sequences. These data may be combined with information on the antibiotic resistance profile of indigenous microbial populations to form a basis for the selection and development of host marker systems. A study of antibiotic resistance within the compost microflora showed that there was a particularly high degree of sensitivity of all groups of thermophilic bacteria to tetracycline (3). Genes encoding tetracycline resistance are now being used in host vector systems to enable the recovery and enumeration of host organisms from compost systems (4). The antibiotic study also revealed that thiostrepton resistance, marker genes for which are being used in mesophilic *Streptomyces* release studies (86, 111, 113), is inappropriate as a primary marker system for compost studies. This is because thiostrepton exhibits poor thermostability at 50°C and therefore has only a limited inhibitory effect on the recovery of indigenous, thiostrepton-sensitive thermophilic streptomycetes at this temperature (3).

RELEASE STUDIES WITH MODEL GRAM-POSITIVE SPECIES

The use of compost for release and recovery experiments has been limited to a few of the predominant species. However, a number of interesting observations have resulted, not least the observation that a character positive for isolation and identification of a target species can be lost on its release into the model microcosm.

Saccharomonospora viridis

The thermophilic actinomycete, *Saccharomonospora viridis*, is being developed as a model host organism for release studies. This organism is indigenous to mushroom compost at concentrations of 10^4 cfu per g (3) and has been isolated from a range of other environments including hay, straw, grain, bagasse, cotton (65), animal faeces (47), sewage sludge compost (79), brown earth soil and upland lake sediment (Amner, unpublished results). One strain has been shown to produce an extracellular xylanase and is of potential economic importance in the removal of xylan from paper pulp (75). *S. viridis* has been selected as a host organism because it can be readily identified on isolation plates by production of a deep green pigment and confirmed microscopically by the characteristic arrangement of single spores on unbranched sporophores of the aerial hyphae. In addition a stable pigmented variant of *S. viridis* is available which is biochemically identical to wild-type strains but produces a deep lilac pigment (77). This strain, BD125, is not isolated from mushroom compost and can therefore be used to differentiate between released (purple) and indigenous (green) strains using pigmentation as a primary marker.

Survival and recovery of *S. viridis* BD125 in mushroom compost has been thoroughly characterized. Detection limits for the strain, originally defined at 10^5 per g fresh compost (3) have been improved through the systematic development of a selective isolation medium, R8 (2). This medium has been successfully used to monitor population changes in both indigenous and released (BD125) *S. viridis* strains in mushroom compost under prolonged laboratory conditions (Fig. 4.3). The introduced strain was found to have no significant effect on either the native population of *S. viridis* or on the total population of thermophilic bacteria, and could be recovered from compost at concentrations of 10^4 g^{-1} after 20 days at 50°C (Fig. 4.3).

Attempts are now being made to introduce genetic markers into *S. viridis* strain BD125 for risk assessment studies. While gene exchange mechanisms

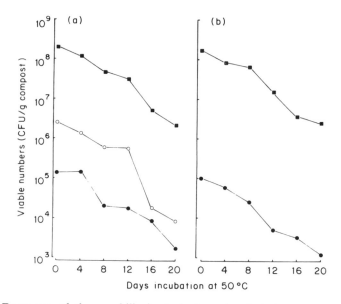

Fig. 4.3 Recovery of thermophilic bacteria from incubated compost microcosms inoculated (a) or uninoculated (b) with *Saccharomonospora viridis* BD125. Total bacteria (■); indigenous *S. viridis* (●); released *S. viridis* BD125 (○). Values are the means of triplicate samples recovered on R8 medium.

have not been characterized for this species, gene transfer in soil is expected to be via conjugal transfer of transmissible plasmids or transduction by actinophage. Phage infection may be of particular importance since *S. viridis* colonies have been isolated from mushroom compost with a pitted aerial mycelium, indicating the presence of phage.

Thermophilic streptomycetes

Streptomycetes are of immense commercial importance due to the large and diverse range of natural products they synthesize, usually as secondary metabolites. It is therefore not surprising that the molecular biology of these organisms is an area of much interest and activity. Because of the rapid advances in recombinant DNA techniques the genes that encode for a number of antibiotic biosynthetic pathways have been clone (17, 29, 51) and genetic manipulation techniques have progressed sufficiently for hybrid antibiotics to have been engineered (52). Work on the fate and survival of genetically manipulated members of this genus is therefore a matter of some urgency, especially since they have an important role in the environment in degrading polymeric materials. Thus their role in the cycling of carbon, which may often be present in relatively recalcitrant polymers, is of great significance in soil mineralization and productivity. Some of the work on survival of streptomycetes in soil is described elsewhere in this volume.

We have used compost as a solid-state matrix and substrate for monitoring the survival of a thermophilic streptomycete. Because compost has both mesophilic and thermophilic phases a streptomycete capable of survival at these temperature extremes has been selected. The species used for these studies is *Streptomyces thermoviolaceus*, which was originally isolated from horse bedding. Other traits of this species advantageous for this work are: a temperature range for growth of 20–57°C; the production of a pigmented antibiotic, granaticin (which is red at acid and blue at alkaline pH), making it easily identifiable on isolation plates from the indigenous streptomycete population within the compost; and a growth physiology of the organism which is relatively well studied compared to other thermophilic species (57,58).

Preliminary work using autoclaved compost showed that *S. thermoviolaceus* could survive in this substrate and recovery of released inocula was high at around 80% down to 10^{-3} cfu per g compost (Fig. 4.4). When *S. thermoviolaceus* was released into non-sterilized raw compost an interesting phenomenon was observed (Fig. 4.4). Up to 2 d after inoculation the streptomycete could be isolated and counted as those colonies producing a blue diffusible pigment. The mean recovery was significantly reduced to 1–2%. After 3 d no pigmented antibiotic-producing colonies could be

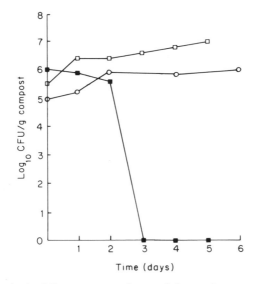

Fig. 4.4 The survival of *Streptomyces thermoviolaceus* in compost. Fresh compost was inoculated with a mycelial suspension of *S. thermoviolaceus* and incubated at 50°C. Total streptomycetes (□); pigment-producing colonies of *S. thermoviolaceus* (■); numbers of *S. thermoviolaceus* colonies after inoculation into sterilized compost (○).

detected. This trait, originally selected as a positive marker, was lost and no pigment was produced even after prolonged incubation. Thus *S. thermoviolaceus* could not be distinguished from indigenous species. When some of the non-pigmented colonies were used as inocula onto agar that contained the same isolation medium the ability to produce the pigmented antibiotic granaticin was regained. Thus compost appeared to transiently affect the physiology of the thermophile and of course meant that antibiotic production on recovery plates was unsuitable as a positive marker for this organism (Featherstone & Edwards, unpublished work). The experiment illustrates the difficulties in extrapolating cell properties observed in laboratory cultures to those manifest in natural environments. The transient loss of a cell property due to passage through a complex natural substrate is therefore significant for risk assessment of a genetically engineered organism.

Bacillus

Release studies involving *Bacillus* spp. have been extremely limited in spite of the industrial importance of these bacteria and the numerous studies of gene transfer processes within members of the genus (55). Mushroom compost contains an abundant source of both mesophilic and thermophilic strains and therefore provides an ideal environment for studying *in situ* gene exchange between these two groups. Transfer of *Bacillus* plasmids from mesophilic to thermophilic hosts (81, 117) and from thermophilic to mesophilic hosts (1, 36, 56, 63) has already been demonstrated *in vitro* but not under environmental conditions.

A number of properties of *Bacillus* make survival studies and predictions of longevity important. The genus contains many commercially important species some of which, such as *B. thuringiensis*, which has insecticidal properties, are destined for deliberate release as bio-control agents. Different species exhibit a range of growth temperatures that span mesophilic to extremely thermophilic conditions; many species, e.g. *B. coagulans*, have growth temperatures (30–60°C) that impinge on both mesophilic and thermophilic temperature domains and therefore have potential for gene transfer between both mesophiles and thermophiles. Even *B. subtilis*, normally considered mesophilic, can also grow at 50°C. Finally, and perhaps more importantly, *Bacillus* species can sporulate to form extremely resistant and long-lived endospores which means that any genetically modified strains have the potential to be extremely long-lived.

Bacillus subtilis containing plasmid pAB224 that encodes tetracycline resistance was used as a model GEM for release studies into compost (4). Only low numbers of indigenous bacteria resistant to tetracycline are found in compost, although several plasmid vectors from thermophilic bacilli encode tetracycline resistance (28). Tetracycline resistant bacteria in compost comprise approximately 10^3 cfu per g compost compared to a total *Bacillus* population of 10^8 cfu per g compost. As a prerequisite to release of *B. subtilis* the small tetracycline-resistant population in compost was in-

Table 4.4 Characteristics of tetracycline-resistant *Bacillus* isolates which contained plasmid DNA (from Ref. 4, with permission).

Strain number	Apparent plasmid size (kb)	Growth at 37°C*		Growth at 65°C*		Starch hydrolysis	Mucoid colony form	Additional drug resistances† detected (μg ml^{-1})			
		NA	NA+Tet	NA	NA+Tet			Kan (100)	Str (50)	Cam (25)	Amp (25)
CB21, 26, 28, 29, 30, 31, 32, 34, 35, 37, 38, 41, 42	10	+	+	−	−	−	−	−	−	−	−
CB9	7	+	+	+	+	−	−	−	+	−	−
CB58	7	+	+	±	±	+	+	−	−	−	+
CB10	>25	+	+	+	+	+	−	−	+	−	−
CB18	>25	+	+	+	−	+	−	−	+	−	−

* Growth was tested in the presence and absence of tetracycline (Tet) at 25μg ml^{-1}.
† Kan, kanamycin; Str, streptomycin; Cam, chloramphenicol; Amp, ampicillin.

vestigated further. Differences in colony morphology were used to select six representative tetracycline-sensitive isolates, and these, plus 52 tetracycline-resistant isolates, were screened for the presence of plasmid DNA. Plasmids could be detected in 17 of the tetracycline-resistant isolates but not in any of the sensitive strains. The plasmid-bearing strains were classified into five groups based on plasmid size, growth at 65°C, starch hydrolysis, mucoidy and antibiotic resistance (Table 4.4). Some homology between plasmid DNA of tetracycline-resistant isolates and pAB224 as well at pTB90 (56) could be demonstrated by Southern hybridization, and the 7 kb and 10 kb plasmids described in Table 4.1 have been used to transform *B. subtilis* to tetracycline resistance.

Survival of *B. subtilis* containing pAB224 (13) could be easily monitored. Only two colony types were detected in the indigenous bacterial populations and these were markedly different to *B. subtilis* colonies. Survival of *B.*

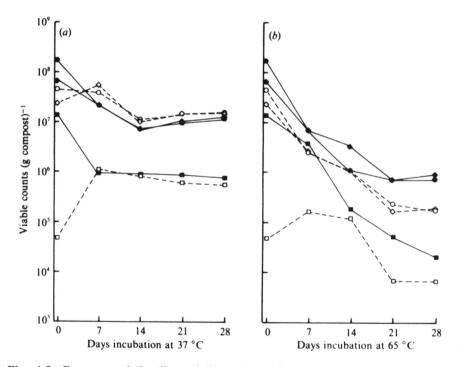

Fig. 4.5 Recovery of *Bacillus subtilis* strain 168 containing pAB224 from, and population dynamics of indigenous bacteria, in fresh compost incubated at (a) 37°C and (b) 65°C. Recovery of the total population from uninoculated control composts on nutrient agar: total counts (◆); spore counts (◇). Recovery of total population from *B. subtilis*-inoculated composts on nutrient agar: total counts (●); spore counts (○). Recovery of tetracycline-resistant population from composts inoculated with *B. subtilis* on nutrient agar containing tetracycline: total counts (■); spore counts (□). All compost samples were recovered and grown at 37°C and values are the means of triplicate determinations (from Ref. 4, with permission).

subtilis in fresh compost was monitored by incubation at 37°C or 65°C (a temperature non-permissive for vègetative growth) and by recovery onto plates of nutrient agar that contained 25 μg ml^{-1} tetracycline and which were incubated at 37°C. At 37°C *B. subtilis* numbers declined during the first two weeks and stabilized thereafter due to sporulation (Fig. 4.5a). At 65°C the population decline was more marked and viable counts in excess of 10^4 cfu g^{-1} were recovered after 28 d.

This experiment revealed a number of important points. First, even a mesophilic *Bacillus* sp. could survive at a non-permissive temperature for growth. Plasmid pAB224 did not appear to be lost from the cells or spores at either incubation temperature. Random selection of *B. subtilis* colonies from non-selective plates also showed no plasmid loss. No *B. subtilis* colonies were detected from uninoculated control composts. Spore counts generally equated with total counts at 37°C but surprisingly were lower at 65°C— possibly due to the formation of immature spores.

This study also showed how an introduced plasmid can survive in a natural population in the *absence* of selection pressure. The long-term survival of plasmids in spores was also demonstrated and thus provides a way forward for further studies on gene exchange between mesophilic and thermophilic populations. The increased longevity of endospore-forming bacteria, even under hostile conditions, has important implications for long-term survival of GEMs in the environment.

In the present release study no gene transfer was demonstrated between released *B. subtilis* and indigenous *Bacillus* populations (4). However, conditions for natural gene transfer between bacilli presumably do exist in compost. Bacilli exhibit natural competence and gene exchange by transformation has previously been demonstrated in soil (44). In addition conjugal gene transfer between *B. subtilis* and *B. cereus* has been described in soil and shown to be increased by the addition of nutrients and the presence of bentonite clay (117). There are also several reports on the isolation and characterization of bacteriophage that infect thermophilic *Bacillus* spp. (54, 91, 94, 112) and these provide the potential for gene transfer by transduction.

SUMMARY

Mushroom compost has been characterized as a model microcosm for GEM risk assessment studies with particular reference to thermophilic Gram-positive species. The microbial population and molecular biology of mushroom compost have aready been evaluated and techniques are currently being developed to improve detection of specific species and enable assessment of gene transfer events and ecological disruptions. Within compost, biological activity is high, nutrients are readily available and the presence of self-replicating DNA units (plasmids) and indigenous bacteriophage, provide favourable conditions for gene exchange. Preliminary release experiments using *B. subtilis* as host organism have shown that introduced plasmids may be maintained for long periods in mushroom compost in the absence of selective pressure and under extreme temperature conditions.

FUTURE PROSPECTS

Microcosm studies are essential for generating data on the behaviour of GEMs prior to their free release into the environment. An increasing number of studies using model systems have clearly demonstrated the potential for gene exchange between introduced bacterial populations. Future studies should be directed towards determining the possibilities of gene transfer to indigenous species and the assessment of disruptions to ecological cycles. Such studies will require the development of increasingly reliable and sensitive methods for detecting released and indigenous microorganisms in the environment. The most promising methods for thermophilic Gram-positive species are expected to involve improved isolation media in combination with hybridization to specific DNA and RNA probes, and ELISA detection of monoclonal and polyclonal antibodies. Emphasis also should be placed on larger scale modelling where the natural environment can be mimicked to a greater degree of accuracy. Scale-up facilities allow environmental parameters to be controlled more rigorously and ensure that engineered inoculants are introduced in a manner realistic of field trial situations.

Traditional composting activities tend to be small scale but are widespread in both domestic and agricultural situations. They may therefore represent a potential sink for GEMs released during agricultural and veterinary practices, and a subsequent source of novel DNA for release into drainage systems and the environment as a whole. Possible environmental disruptions are increased by the ability of many species in compost to colonize a range of environmental substrates and to survive for extended periods as spores. Composting systems are therefore expected to continue to play a role in risk assessment studies in the future.

ACKNOWLEDGEMENT

We wish to acknowledge the financial support of the NERC for the experimental work described in this chapter.

REFERENCES

1. Aiba, S., Kitai, K. & Imanaka, T. (1983) *Applied and Environmental Microbiology* **46**, 1059–1065.

2. Amner, W., Edwards, C. & McCarthy, A.J (1989) *Applied and Environmental Microbiology* **55**, 2669–2674.
3. Amner, W., McCarthy, A.J. & Edwards, C. (1988) *Applied and Environmental Microbiology* **54**, 3107–3112.
4. Amner, W., McCarthy, A.J. & Edwards, C. (1990) *Journal of General Microbiology* **137**, 1931–1937.
5. Andersen, A.A., (1958) *Journal of Bacteriology* **76**, 471–484.
6. Athalye, M., Lacey, J. & Goodfellow, M. (1981) *Journal of Applied Bacteriology* **51**, 289–297.
7. Atkey, P.T. & Wood, D.A. (1983) *Journal of Applied Bacteriology* **55**, 293–304.
8. Bagstam, G. (1979) *European Journal of Applied Microbiology and Biotechnology* **6**, 279–288.
9. Bale, M.J., Day, M.J., & Fry, J.C. (1988) *Applied and Environmental Microbiology* **54**, 2756–2758.
10. Barkay, T., Leibert, C. & Gillman, M. (1989) *Applied and Environmental Microbiology* **55**, 1574-1577.
11. Bentjen, S.A., Fredrickson, J.K., Van Voris, P. & Li, S.W., (1989) *Applied and Environmental Microbiology* **55**, 198–202.
12. Berg, J.D. & Fiksdal, L. (1988) *Applied and Environmental Microbiology* **54**, 2118–2122.
13. Bingham, A.H.A., Bruton, C.J. & Atkinson, T. (1980) *Journal of General Microbiology* **199**, 109–115.
14. Bleakley, B.H. & Crawford, D.L. (1989) *Canadian Journal of Microbiology* **35**, 544–549.
15. Brooks, R.C., Fermor, T.R. & McCarthy, A.J. (1990) *Journal of Scientific Food and Agriculture* **50**, 132–133.
16. Carnes, R.A. & Lossin, R.D. (1970) *Composting Science* **11**, 18–21.
17. Chater, K.F., Hopwood, D.A., Kieser, T. & Thompson, C.J. (1982) *Current Topics in Microbiology and Immunology* **97**, 69–95.
18. Chaundhry, G.R., Toranzos, G.A. & Bhatti, A.R. (1989) *Applied and Environmental Microbiology* **55**, 1301–1304.
19. Cresswell, N., Herron, P.R., Saunders, V.A. & Wellington E.M.H (1992) *Journal of General Microbiology* **138**, 659–666.
20. Cross, T. (1968) *Journal of Applied Bacteriology* **31**, 36–53.
21. Dauenhauer, S.A., Hull, R.A. & Williams, R.P. (1984) *Journal of Bacteriology* **158**, 1128–1132.
22. Derikx, P.J.L., de Jong, G.A.H., Op den Camp, H.J.M., van der Drift, C., van Griensven, L.J.L.D. & Vogels, G.D. (1989) *FEMS Microbiology Ecology* **62**, 251–258.
23. Derikx, P.J.L., Op den Camp, H.J.M., van der Drift, C., van Griensven, L.J.L.D. & Vogels, G.D. (1990) *Applied and Environmental Microbiology* **56**, 176–186.
24. De Rossi, E., Brigidi, P., Riccardi, G. & Matteuzzi, D. (1989) *Current Microbiology* **19**, 13–19.
25. Devanas, M.A. & Stotzky, G. (1986) *Current Microbiology* **13**, 279–283.
26. Drahos, D.J., Hemmings, B.C. & McPherson, S. (1986) *Biotechnology* **4**, 439–444.
27. Dutkiewicz, J., Olenchock, S.A., Sorenson, W.G., Gerencser, V.F., May, J.J., Pratt, D.S. & Robinson, V.A. (1989) *Applied and Environmental Microbiology* **55**, 1093–1099.

28. Edwards, C (1991) In *Microbiology of Extreme Environments*, Ed. Edwards, C. Open University Press, Milton Keynes, pp. 1–32.
29. Feitelson, J.S., Malpartida, F. & Hopwood, D.A. (1985) *Journal of General Microbiology* **131**, 2431–2441.
30. Fermor, T.R., Randle, P.E. & Smith J.F. (1985) In *Compost as a Substrate and its Preparation*, Flegg, P.B., Spencer, D.M. & Wood D.A., Eds John Wiley & Sons, Chichester, pp. 81–109.
31. Fermor, T.R., Smith, J.F. & Spencer, D.M. (1979) *Journal of Horticultural Science* **54**, 137–147.
32. Fiskel, J. & Covello, V.T. (Eds) (1986) *Biotechnology Risk Assessment. Issues and Methods for Environmental Introductions* Pergamon Press, New York.
33. Finstein, M.S. (1989) *ASM News* **55**, 599–602.
34. Ford, S. & Olsen, B.H. (1988). *Advances in Microbial Ecology* **10**, 45–79.
35. Fredrickson, J.K., Bentjen, S.A., Bolton, H., Li, S.W. & van Voris, P. (1989) *Canadian Journal of Microbiology* **35**, 867–873.
36. Fujii, M., Takagg, M., Imanaka, T. & Aiba, S. (1983) *Journal of Bacteriology* **154**, 831–837.
37. Gealt, M.A., Chai, M.D., Alpert, K.B. & Boyer, J.C. (1985) *Applied and Environmental Microbiology* **49**, 836–841.
38. Geisser, M., Tischendorf, G.W. & Stoffler, G. (1973) *Molecular and General Genetics* **127**, 129–145.
39. Germida, J.J. & Khachatourians, G.G. (1988) *Canadian Journal of Microbiology* **34**, 190–193.
40. Godden, B., Penninckx, M., Pierard, A. & Lannoye, R. (1983) *Applied Microbiology and Biotechnology* **17**, 306–310.
41. Golovleva, L.A., Pertsova, R.N., Boronion, A.M., Travkin, V.M & Kozlovsky, S.A (1988) *Applied and Environmental Microbiology* **54**, 1587–1590.
42. Goodfellow, M., Lacey, J. & Todd, C. (1987) *Journal of General Microbiology* **133**, 3135–3149.
43. Goodfellow, M., & Williams, E. (1986) *Biotechnology and Genetic Engineering Reviews* **4**, 213–262.
44. Graham, J.B. & Istock, C.A. (1978) *Molecular and General Genetics* **166**, 287–290.
45. Gregory, P.H. & Lacey, M.E. (1963) *Journal of General Microbiology* **30**, 75–88.
46. Griener-Mai, E., Kroppensteddt, R.M., Korn-Wendisch, F. & Kutzner, H.J. (1987) *Systematics and Applied Microbiology* **9**, 97–109.
47. Hayashida, S., Nanri, N., Teramoto, Y., Nishimoto, T., Ohta, K. & Miyaguchi, M. (1988) *Applied and Environmental Microbiology* **54**, 2058–2063.
48. Heitkamp, M.A., Freeman, J.P. & Cerniglia, C.E. (1987) *Applied and Environmental Microbiology* **53**, 129–136.
49. Hogan, J.A., Miller, F.C. & Finstein, M.S. (1989) *Applied and Environmental Microbiology* **55**, 1082–1092.
50. Holben, W.E., Jansson, J.K., Chelm, B.K. & Tiedje, J.M. (1988) *Applied and Environmental Microbiology* **54**, 703–711.
51. Hopwood, D.A., Bibb, M.J., Bruton, C.J., Feitelson, J.S. & Gil, J.A. (1983) *Trends in Biotechnology* **1**, 42–48.
52. Hopwood, D.A., Malpartida, F., Kieser, H.M., Ikeda, H., Duncan, J., Fujii, I. et al. (1985) *Nature* **314**, 642–644.

53. Hoshino, T., Ikeda, T., Narushima, H. & Tomizuka, N. (1985) *Canadian Journal of Microbiology* **31**, 339–345.
54. Humbert, R.D. & Fields, M.L. (1972) *Journal of Virology* **9**, 397–398.
55. Imanaka, T. & Aiba, S. (1986) In *Thermophiles: General Molecular and Applied Microbiology*, Ed. Brook, T.D., John Wiley & Sons, New York, pp. 159–179.
56. Imanaka, T., Fujii, M., Aramori, I. & Aiba, S. (1982) *Journal of Bacteriology* **149**, 824–830.
57. James, P.D.A. & Edwards, C. (1988) *FEMS Microbiology Letters* **52**, 1–6.
58. James, P.D.A. & Edwards, C. (1989) *Journal of General Microbiology* **135**, 1997–2003.
59. Kemp, H.A., Archer, D.B. & Morgan, M.R.A. (1988) *Applied and Environmental Microbiology* **54**, 1003–1008.
60. Krasovsky, V.N. & Stotzky, G. (1987) *Soil Biology and Biochemistry* **19**, 631–638.
61. Kuckek, K. & Mordarsski, M. (1989) *FEMS Microbiology Letters* **61**, 257–260.
62. Kuriki, T., Park, J., Okada, S. & Imanaka, T. (1988) *Applied and Environmental Microbiology* **54**, 2881–2883.
63. Kurup, V.P. (1984) In *Thermophilic Actinomycetes: Their Role in Hypersensitivity Pneumonitis,* Eds Ortiz-ortiz, L., Bojalil, L.F., & Takoleff, V. Academic Press, London, pp. 145–159.
64. Lacey, J. (1973) In *Actinomycetes in Soils, Composts and Fodders*, Eds Skykes, G., & Skinner, F.A. Academic Press, London, pp. 231–251.
65. Lacey, J (1978) In *Ecology of Actinomycetes in Fodders and Related Substrates*, Eds Mordarski, M., Kurylowicz, W. & Jeljaszewicz, J. Gustav Fisher Verlag, Stuttgart, pp. 161–170.
66. Lacey, J. (1981) In *Actinomycetes*, Eds Schaal, K.P., & Pulverer, G. Gustav Fisher Verlag, Stuttgart, pp. 243–250.
67. Lacey, J. & Dutkiewicz, J. (1976) *Journal of Applied Bacteriology* **41**, 13–27.
68. Lacey, J. & Dutkiewicz, J. (1976) *Journal of Applied Bacteriology* **41**, 315–319.
69. Lawrence, H.M., Merivuori, H., Sands, J.A. & Pidcock, K.A. (1986) *Applied and Environmental Microbiology* **52**, 631–636.
70. Lloyd, D. & Scott, R.I. (1983) *Journal of Microbiological Methods* **1**, 313–328.
71. Mancini, P., Fertels, S., Nave, D. & Gealt, M.A. (1987) *Applied and Environmental Microbiology* **53**, 665–671.
72. Martenssen, A.M. (1990) *Canadian Journal of Microbiology* **36**, 136–139.
73. Mayfield, C.I., Williams, S.T., Ruddick, S.M. & Hatfield, H.L (1972) *Soil Biology and Biochemistry* **4**, 79–91.
74. McCarthy, A.J. (1985) *Frontiers of Applied Microbiology* **1**, 1–14.
75. McCarthy, A.J. (1987) *FEMS Microbiology Reviews* **46**, 145–163.
76. McCarthy, A.J. & Cross, T. (1981) *Journal of Applied Bacteriology* **51**, 299–302.
77. McCarthy, A.J. & Cross, T. (1984) *Journal of General Microbiology* **130**, 5–25.
78. McClure, N.C., Weightman, A.J. & Fry, J.C. (1989) *Applied and Environmental Microbiology* **55**, 2627–2634.
79. Millner, P.D. (1982) *Developmental and Industrial Microbiology* **23**, 61–78.
80. Morel, J.L., Bitton, G., Chaudhry, G.R. & Awong, J. (1989) *Current Microbiology* **18**, 355–360.
81. Nakamura, K. & Imanka, T. (1989) *Applied and Environmental Microbiology* **55**, 3208–3213.
82. Ochi, K. (1989) *Journal of General Microbiology* **135**, 2635–2642.

83. Pidcock, K.A., Montenecourt, B.S. & Sands, J.A. *FEMS Microbiology Letters* **55**, 349–352.
84. Polack, J. & Novisk, R.P. (1982) *Plasmid* **7**, 152–162.
85. Rafaii, F. & Crawford, D.L. (1988) *Applied and Environmental Microbiology* **54**, 1334–1340.
86. Rafaii, F. & Crawford, D.L. (1989) *Current Microbiology* **19**, 115–121.
87. Richaume, A., Angle, J.S. & Sadowsky, M.J. (1989) *Applied and Environmental Microbiology* **55**, 1730–1734.
88. Rissler, J.F. (1984) *Recombinant DNA Technical Bulletin* **7**, 20–30.
89. Rochelle, P.A., Fry, J.C. & Day, M.J. (1989) *FEMS Microbiology Ecology* **6**, 127–136.
90. Rozak, D.B., & Colwell, R.R. (1987) *Applied and Environmenal Microbiology* **53**, 2889–2983.
91. Saunders, G.F. & Campbell, L.L. (1966) *Journal of Bacteriology* **91**, 340–348.
92. Saye, D.J., Ogunseitan, O.A., Sayler, G.S. & Miller, R.V. (1990) *Applied and Environmental Microbiology* **56**, 140–145.
93. Scanferlato, V.S., Orvos, D.R., Cairns, J. & Lacy, G.H. (1989) *Applied and Environmental Microbiology* **55**, 1477–1482.
94. Sharp, R.J., Ahmad, S.I., Munster, A., Dowsett, B. & Atkinson, T (1986) *Journal of General Microbiology* **132**, 1709–1722.
95. Shaw J.J. & Kado, C.I. (1986) *Biotechnology* **4**, 560–564.
96. Smith, J.F. & Spencer, D.M. (1976) *Scientia Horticulturae* **5**, 23–31.
97. Smith, J.F. & Spencer, D.M. (1977) *Scientia Horticulturae* **7**, 197–205.
98. Smith, J.F. (1983) *Scientia Horticulturae* **19**, 65–78.
99. Stanek, M. (1972) *Mushroom Science* **8**, 797–811.
100. Strom, P.F. (1985) *Applied and Environmental Microbiology* **50**, 899–905.
101. Strom, P.F. (1985) *Applied and Environmental Microbiology* **50**, 906–913.
102. Stutzenberger, F.J. (1971) *Applied Microbiology* **22**, 147–152.
103. Treuhaft, M.W. & Arden Jones, M.P. (1982) *Journal of Clinical Microbiology* **16**, 995–999.
104. Trevors, J.T. & Berg, G. (1989) *Systematics and Applied Microbiology* **11**, 223–227.
105. Trevors, J.T. & Van Elsas, J.D. (1989) *Canadian Journal of Microbiology* **35**, 895–902.
106. Trevors, J.T., Van Elsas, J.D., Starodub, M.E. & van Overbeek, L.S (1989) *Canadian Journal of Microbiology* **35**, 675–680.
107. Van Elsas, J.D., Govaert, J.M. & van Veen, J.A. (1987) *Soil Biology and Biochemistry* **19**, 639–647.
108. Van Elsas, J.D., Nikkel, M. & van Overbeek, L.S. (1989) *Current Microbiology* **19**, 375–381.
109. Van Elsas, J.D., Trevors, J.T. & Starodub, M.E. (1988) In *Risk Assessment for Deliberate Release*, Ed. Klingmuller, W. Springer-Verlag, Berlin, pp. 89–99.
110. Walter, M.V., Porteous, L.A. & Seidler, R.J. (1989) *Current Microbiology* **19**, 365–370.
111. Wang, Z., Crawford, D.L., Pometto, A.L. & Rafii, F. (1989) *Canadian Journal of Microbiology* **35**, 535–543.
112. Welker, N.E., (1988) *Journal of Bacteriology* **170**, 3731–3764.
113. Wellington, E.M.H., Saunders, V.A., Cresswell, N. & Wipat, A. (1988) In *Biology of Actinomycetes '88,* Eds Okami, Y, Beppu, T. & Ogawara, H. Japan Scientific Societies Press, Tokyo, pp. 300–305.

114. Williams, S.T., Shameemullah, M., Watson, E.T. & Mayfield, C.I. (1972) *Soil Biology and Biochemistry* **4**, 215–225.
115. Williams, S.T. & Vickers, J.C. (1988) In *Biology of Actinomycetes '88,* Eds Okami, Y., Beppu, T. & Ogawara, H. Japan Scientific Societies Press, Tokyo pp. 265–270.
116. Winstanley, C., Morgan, J.A.W., Pickup, R.W., Jones, G.J. & Saunders, J.R (1989) *Applied and Environmental Microbiology* **55**, 771–777.
117. Wu, L. & Welker, N.E. (1989) *Journal of General Microbiology* **135**, 1315–1324.
118. Yeung, K.A., Schell, M.A. & Hartel, P.G. (1989) *Applied and Environmental Microbiology* **55**, 3243–3246.
119. Zeph, L.R. & Stotzky, G. (1989) *Applied and Environmental Microbiology* **54**, 1731–1737.

Chapter 5

Extraction, Detection and Identification of Genetically Engineered Microorganisms from Soils

A.G. O'Donnell[1] and D.W. Hopkins[2]
[1]Department of Agricultural and Environmental Science,
University of Newcastle upon Tyne
[2]Department of Biological Sciences, University of Dundee

Introduction

The ability to monitor reliably and reproducibly, genetically engineered microorganisms (GEMs) in natural ecosystems is considered vital in assessing their environmental impact. Such data can be obtained from an analysis of either the organism's phenotype or by direct detection of its genetic composition. However, the ability to determine whether a given microorganism is present in the environment is a difficult task (31). Phenotypic approaches measure 'the perceptible properties or other characteristics of an organism resulting from the interaction of its genetic constitution and the environment'; whereas a genotypic approach detects the 'genetic constitution of an organism,

Monitoring Genetically Manipulated Microorganisms in the Environment. Edited by C. Edwards
Published 1993 John Wiley & Sons Ltd. © 1993 A.G. O'Donnell and D.W. Hopkins

as distinct from its physical and functional appearance' (31). The success of both approaches is largely dependent on the complexity of the environment, with properties such as the density, distribution and diversity of the population and environmental factors such as temperature, pH, redox and surface sorption characteristics having an important role. Detecting a target organism also depends on its ability to establish itself or its genetic trait in the affected environment. This is likely to depend on the size and activity of the initial inoculum and on the effectiveness with which it is dispersed (31).

Much of the work designed to provide better risk assessment has concentrated on the development of more sensitive, specific and reproducible methods for detection, but in using these techniques it is important to ensure that experiments are designed in such a way that the results obtained are of relevance to the situations and the environments in which the release experiments are likely to take place. This makes sampling (35) and an understanding of the environment as important as the detection method used. Amongst natural habitats, soil is perhaps the most complex and it may be necessary to sample extensively to ensure that a sufficiently representative sample is obtained (88). For those detection techniques based on the extraction of microorganisms from soil it is critical to determine the efficiency with which microorganisms are extracted although, as discussed below, this can be very difficult to quantify accurately.

In this chapter, the nature of soil as a habitat for microbial growth, genetic exchange and survival is briefly described. The particular problems that soil presents in sampling and monitoring microbial populations are highlighted and the general approaches and specific attempts to overcome them described. There have been a number of excellent reviews and descriptions of soil as a habitat for the growth, genetic exchange and survival of microorganisms (41, 47, 113); it is not the intention of this chapter to review these, but the reader is referred to them for a wider coverage of the subject.

SOIL AS A HABITAT FOR MICROORGANISMS

Soil constituents

Soil is composed of a heterogeneous mixture of solid, liquid and gaseous phases, all of which vary both spatially and temporally. The solid materials can be divided into mineral and organic components, with the mineral component further divided into sand, silt and clay fractions.

The mineral component of soil

Sand particles (equivalent spherical diameter, ESD>212 μm) are either primary minerals or insoluble secondary minerals. Since they have a small specific area and surface reactivity and are often coated with other materials, they have little direct effect on soil microorganisms (112). Their influence is confined to indirect effects arising from the soil texture. Similarly, the silt fraction (ESD 63–212 μm), which is composed of many of the same minerals as the sand fraction, has an indirect effect on soil microorganisms. Nevertheless, when soils are dispersed, a large proportion of the soil microbial biomass is associated with the fine silt-sized fraction (2), because of the large amount of soil organic matter (SOM) that is also present in this fraction (63).

Many particles in the clay fraction (>2 μm) of soil are colloidal and interact significantly with soil microorganisms. Clay particles are principally crystalline secondary minerals derived from the weathering of primary minerals, but the clay fraction also contains accessory minerals, salts and amorphous and crystalline metal oxides. The crystalline clay minerals are usually either 1:1 (e.g. kaolinite) or 2:1 (e.g. montmorillonite and illite) layer lattice structures consisting of repeated lamellae of silica (silicon oxide tetrahedral crystal) and gibbsite (aluminium hydroxide octahedral crystal) sheets. Isomorphous substitution in the crystal structures by elements with the same hydration number but different valency leads to charge imbalances within clay minerals. The most common substitutions are Mg^{2+} for Al^{3+} and Al^{3+} for Si^{4+}, which lead to an excess negative charge that is fixed in both amount and space and that has to be neutralized by attraction of exchangeable cations to the surface of the lattice. In addition, there are positive charges present at the exposed edges of lamellae, but since these are outnumbered by the negative charges, clay minerals have a characteristic cation exchange capacity (CEC).

Amorphous and crystalline free oxides of, for example, silicon, iron and aluminium have two important roles. Firstly they form coatings on the surface of other soil particles and their influence may thus be disproportionate to their presence by weight. Second, they possess pH dependent charge, such that as the pH increases, the amount of positive charge increases with a point of zero charge (PZC) in the range pH 3–10 (90).

Soil organic matter

Mineral soils usually contain less than about 5% organic matter in the surface horizons. Organic soils, by contrast, can be composed almost entirely of organic matter. Soil organic matter (SOM) is a mixture of relatively undecomposed particulate material of either plant or animal origin; well-defined decomposition products such as proteins and carbohydrates; and complex, poorly-defined products of microbial degradation and synthesis collectively known as humic substances. Humic substances are heterogeneous, polymeric, organic materials containing mainly carbon and hydrogen, which possess various functional groups (107). In soil much of the humic material is associated with the clay fraction in the form of organo-mineral complexes and may result in only a small proportion of the mineral surface being exposed (119). Humic polymers are flexible, swell on hydration, have considerable internal surface area and carry a variable charge. This charge arises mainly from the dissociation of carboxyl groups to the carboxylate anion with increasing soil pH and can result in a greater CEC than that of clay minerals (120). SOM is the principal source of metabolizable carbon for many heterotrophic soil microorganisms and sites containing such organic matter are often more heavily colonized by microorganisms than the bulk soil (40,128).

Structured soil

Aggregates are the normal structural units of soil (Fig. 5.2A). They can range from a few micrometres to several centimetres in diameter. Various processes, including flocculation of clay particles in the presence of di- and poly-valent cations, binding action of organic polymers (20), cementing action of metal oxides, adhesive extracellular polymers of some microorganisms (67) and enmeshing by fungal hyphae (75) contribute to either the formation or stability of soil aggregates. The continuous system of pores running through and between aggregates provides routes for water and gas movement into and out of the soil. The surfaces of the pores and voids are usually covered in a film of water, which even under particularly dry conditions is tenaciously held by surface tension. It is this network of pores and voids on soil aggregates that provides the habitat in which microorganisms reside (47).

Depending on the sizes of the pores they will be filled with either air or water. At low, more negative, moisture potentials only the narrowest pores will be water-filled, whereas at higher moisture potentials larger pores will become water-filled (89). Within the typical range of moisture potentials the range of pore sizes that are water-filled is approximately from 0.1 to 20 μm. Microorganisms resident in the larger pores are likely to be subject to greater fluctuations in available water than those in smaller pores (43). This influences the type of organism that can survive in the differently sized pores (63) and also the movement and colonization by microorganisms in soils. Fungi, because of their ability to translocate water and grow directionally can, unlike non-filamentous microorganisms, cross air-filled spaces.

SOIL–MICROBE INTERACTIONS

Retention of microorganisms by soil aggregates

There is good empirical evidence that microorganisms are retained by entrapment within soil aggregates (47) and there are numerous examples of spatial heterogeneity in the distribution of microorganisms within aggregates (32,52,61). There are both quantitative and qualitative differences in the numbers, type and responses of microbial cells determined by the size and location of the pores in which the cells reside. For example, it has been shown that protozoans (97%) and fungi (80–90%) were mainly located on the outer parts of aggregates, whereas bacteria (97%) existed predominantly in the inner parts (46). Bacteria occupying the outer surfaces of aggregates were predominantly Gram-positive spore-forming organisms, whereas the interior of the aggregates contained a significantly greater proportion of Gram-negative vegetative cells. This is consistent with the micro-environment at the inner parts of aggregates being less prone to intermittent desiccation conditions and, therefore, leading to an increase in the survival chances of non-spore-forming bacteria (62).

Reactions at surfaces

Much of our knowledge about the interactions between soil microorganisms and soil colloids is based on studies using pure cultures of microorganisms and homoionic clays. Whilst there is good evidence that the presence of clay minerals affects the survival of microorganisms by, for example, providing physical protection from predation by protozoa (1, 49, 123) and parasitism by *Bdellovibrio* spp. (60) or by indirect effects on the physico-chemical environment (108, 109, 110) it remains unclear how significant such interactions are in natural soils. That aggregate dispersion and formation, pore collapse and rearrangement, and water movement occur frequently in soil due to natural temperature and moisture content fluctuations, mesofaunal action, plant root growth and man's activities suggests that mechanisms other than retention or entrapment within aggregates are involved in the retention of microorganisms in soil (114).

Interactions between microorganisms and soil particles are generally governed by the colloidal properties of their surfaces. The ionogenic groups at the outer surface of the bacterial envelope in Gram-positive bacteria are primarly phosphate ester and carboxyl groups of teichoic and teichuronic acids, respectively. Since the pK_a of these groups is below the normal range of soil pH values, Gram-positive bacteria tend to carry a negative surface charge (4, 12, 64). The amphoteric nature of proteins in the Gram-negative bacterial envelope enables the cell surface to carry both positive and negative charge depending on the ionization of amino and carboxyl groups (71), but the balance is usually towards a negative charge (11, 72). Since most soil colloids also carry a net negative surface charge, the electrokinetic potential (EKP or zeta potential) that exists between two like charges must be reduced if they are to come into contact. To understand how this might occur it is necessary to understand the distribution of ions at colloidal surfaces.

The DLVO—Derajagium & Landau, 1941 (23); Verwey & Overbeek, 1948 (124)—theory of attraction between two like-charged colloids relies on there being a diffuse double layer (DDL) of counter-ions in solution at the charged surfaces (96). The DDL is a layer of counter-ions which form a gradient of declining concentration away from the surface. The DDL forms as a result of the opposing forces of, on the one hand van der Waals forces and electrostatic attraction, and on the other, of thermal dispersion and electrostatic repulsion. The EKP is reduced at high electrolyte concentration in the bulk phase because the extent of the DDL is decreased, so that under these conditions two like-charged colloids can approach each other more closely. At an intermediate electrolyte concentration there is a specific distance between two like-charged colloidal surfaces at which attraction occurs. Although the DDL may in theory extend for between 2 and 40 nm from the colloid surface, in natural soil the water content is often depleted to such an extent that a truncated DDL forms (19). This means that the DLVO theory may only apply to soil colloids and microorganisms dispersed in aqueous suspension (66, 112).

Many studies with homoionic clays and microorganisms have indicated the potential for many types of surface interactions in soil (72, 112). Van der Waals forces, hydrogen bonding, water bridging and coordination complexes may be the most significant processes leading to equilibrium adsorption of the cells to soil colloids (111). The existence of an equilibrium adsorption process is important in attempts to remove bacteria from soil colloids, because it implies that the process is reversible. Bonding, on the other hand, requires that chemical bonds are formed between the bacteria and soil particle. Effectively permanent bonding could also result from the attachment of cells to soil colloids by a number of physical forces, the sum of which exceeded any dissociative forces—the so-called zipper effect (111). Whilst microorganisms in soil can produce adhesive polymers that enable them to adhere to solid surfaces (72), it is unclear whether these are capable of overcoming the EKP or are even produced universally under the nutritionally poor conditions which may occur in soil (114).

Uncertainty over the processes involved in the physical interactions between microorganisms and soil particles, as outlined above, has generally led to difficulties in developing specific methods for quantitatively and representative sampling the soil microfloras. This is reflected in the empirical nature of many of the extraction procedures that are used to prepare samples of microorganisms from soil.

EXTRACTION

Need for extraction and sampling

Since, in any risk assessment the possibility of any transfer event or the long-term survival of GEMs must be taken into account (12), it is essential to ensure that samples used for detecting the introduced genes or genotypes are representative of the soil microflora. It is not, for example, sufficient simply to consider those organisms most readily removed from the soil because, as has been outlined above, microorganisms are not uniformly distributed in soil and are probably not, therefore, uniformly retained by soil.

Various approaches, both rational and empirical, have been developed for sampling soil microorganisms (Table 5.1). Each of these approaches involves steps which can be broadly divided into dispersion, dissociation, separation and purification. It is frequently not possible to distinguish between treatments which lead to the dispersion of soil aggregates and those causing dissociation of soil particles and microorganisms since neither procedure is simple nor independent. Complete soil dispersion requires the destruction of soil aggregates thereby exposing those organisms present inside aggregates. Once exposed, these organisms can then be dissociated from colloidal surfaces. An impure extract can then be obtained by separating the larger and more dense particles from suspension. Further purification can then be attempted. Inevitably, however, in the development of any sampling regime it is necessary to consider the damage caused to, or loss of, the micro-organisms during extraction. This must be offset against the quantity and purity of the sample required.

Dispersion and dissociation

It can be argued that representative samples from soil can only be achieved if the soil is completely dispersed. Where less than 100% soil dispersion is achieved it is more likely that all aggregates are incompletely dispersed rather than some of the aggregates being completely dispersed. Since there are qualitative differences in the distribution of microorganisms in aggregates, incomplete soil dispersion will, inevitably, lead to sites on the insides of soil aggregates remaining unsampled.

Various methods are available for dispersing mineral soils. However, complete and rapid dispersion of all soil colloids by high-power ultrasonic irradiation (42) requires more energy than that necessary to destroy bacterial cells (94). Consequently, the use of low-energy ultrasonic baths is preferable (57, 58, 77).

Mechanical homogenization (9, 29) has been shown to be more efficient than simple shaking at releasing cells from soils (59) and to be particularly valuable in dispersing organic soils (54). This is because the stability of a highly organic soil is more dependent on the integral strength of the particulate matter than on the association of particles (115). Nevertheless, dispersion by shaking with various aqueous solutions such as sodium pyro-phosphate, diluted Winogradsky salt solution, Tris buffer and sodium hexa-metaphosphate (5, 9, 29, 77) have all been used. Most of these reagents work by sequestering divalent cations, particularly calcium. Macdonald (68) has suggested using chelating ion-exchange resin as a sink for di- and polyvalent cations to disperse clay particles (26). This procedure (68), which also involves the anionic detergent sodium cholate, was found to be the most effective of the single-step dispersion methods tested on soils of differing textural class (53).

The aggregate stability conferred by organic matter can be partially overcome using mild detergents. However, in the majority of studies in which detergents have been used, the emphasis has been on reducing the potential adhesive action of extracellular polymers (68, 100). Several different deter-gents have been used, but since many are known to be toxic to micro-organisms (27, 44, 125) a full evaluation is necessary prior to use. Sodium cholate (0.1% w/v) has been proposed as, although not demonstrated to be,

Table 5.1 Summary of some approaches to extracting microorganisms from soil.

Approach	Target organisms	Quoted efficiency*	Biomass estimate	Soil description†	Reference
Homogenization and ultrasonication in sodium pyrophosphate used to release cells. Cells collected by differential centrifugation	Indigenous microorganisms	NV		Silty clay loam	9
	Added *Arthrobacter globiformis*	77[a]	Viable (plate) count	Silty clay loam	10
Repeated (3 times) homogenization in Winogradsky's salt solution used to release cells. Cells were collected after each stage of homogenization by differential centrifugation.	Indigenous bacteria	82[a]	"	Sub-alpine peat	29
		52[a]	"	Peaty podzol	29
		51[a]	"	Peaty heathland soil	29
		35[a]	"	Loamy soil	51
		27[a]	"	Sandy loam	106
As above, but repeated 8 times	"	71[a]	"	Clay loam	5
Shaking in detergent used to release cells which were separated by sedimentaton under gravity for 5 minutes	Added bacteria (*Flavobacterium* sp. *Aeromonas* sp. *Escherichia* sp.)	NV		Marine sludge and clay	100
Shaking in Tris buffer (pH 7.5) used to release cells which were then separated by sedimentation under gravity for 15 minutes	Added bacteria (*Alcaligenes eutrophus*)	16[b]	Viable (plate) count	Meadow chernozem	77
Ultrasonication for 30 seconds (150 W) in Tris buffer used to release cells from soil	Indigenous microorganisms	NV		Silt loam and sandy loam	94

continued on page 119

Table 5.1 *(continued).*

Approach	Target organisms	Quoted efficiency*	Biomass estimate	Soil description†	Reference
Gentle homogenizaton in Tris buffer used to release cells which were collected by rate-zonal density-gradient centrifugation	Indigenous non-filamentous microorganisms	10–20[a]	Direct microscopic count	Silty clay loam	73
Shaking in detergent and chelating exchange resin used to release cells which were collected by elutriation and density-gradient centrifugation	"	NV		Clay loam	68, 69
Repeated shaking with detergent and chelating exchange resins used to release spores which were collected after each stage by low-speed centrifugation	Added *Streptomyces lividans* spores	NV		Sandy loam	48

*NV Indicates that no comparable efficiency values were calculated by the authors.
[a] Indicates that recovery efficiency was expressed relative to the biomass in the extract and residue or whole soil.
[b] Indicates that recovery efficiency was expressed relative to the total amount of added biomass.
† Soil descriptions do not necessarily follow any recognized classification system. They have been extracted from the source references.

a non-toxic detergent which avoids excessive frothing (68). Sodium cholate has been used to some effect in subsequent studies (44, 54).

Separation

The fundamental problem in the separation of the soil biomass from soil particles is that the ranges of their sizes and sedimentation rates under gravity overlap. Consequently, separation procedures such as sieving, differential centrifugation and elutriation result in the removal of only the largest, most dense or fastest sedimenting soil particles.

Sieving has been used to remove particles greater than 30 μm from soil extracts (68) and, for example, in the concentration of soil diatoms (101). However, if the soil is incompletely dispersed, sieving will remove much of the biomass trapped within stable aggregates or particulate matter.

The elutriation principle (39) has been used to separate organisms from soil particles in a number of studies (65, 71, 87, 102, 121, 122). Elutriation (like centrifugation) separates particles on the basis of their sedimentation velocity as defined by Stokes' law:

$$v = \frac{2r^2g(\varrho_p - \varrho_m)}{9\eta\theta}$$

where v = sedimentation velocity (cm s^{-1}); r = particle radius (cm); g = acceleration due to gravity (981 cm s^{-2}); ϱ_p = particle density (g cm^{-3}); ϱ_m = medium density (g cm^{-3}); η = kinematic viscosity (0.01 cm g^{-1}); θ = coefficient of form resistance (74).

Since r is raised to the second power and ϱ_p is not, the value of r has a greater effect on v. The range of values of r for naturally isolated soil bacteria extends below 0.2 μm (7), which is well within the clay particle size range (<2 μm). Values of ϱ_p for microorganisms are less than 1.3 g cm^{-3} (6,29), compared with 2.6 g cm^{-3} for soil mineral particles (based on theoretical value for kaolinite; Ref. 45). This means that the property of microorganisms and soil particles which differs most is least influential in determining sedimentation velocity. There is, therefore, considerable overlap in the ranges of sedimentation velocities of the two populations of particles (54).

Although elutriation gave apparently high biomass recoveries from soil dispersed using sodium cholate and chelating ion-exchange resin (54), there are a number of disadvantages with this method. Separation by elutriation has proved impractical since the flow rates required were too slow and the sample obtained had a large volume and was very dilute (54). Concentration of the elutriate by hollow fibre ultrafiltration has been attempted (68) but it is likely that the attrition during this process would be excessively damaging to microorganisms.

Differential centrifugation is a more widely used alternative, despite the fact that the separation of particles is also governed by Stokes' law. It has been suggested that cells co-sediment with soil particles during centrifugation (68), but it is not clear whether this is due to incomplete aggregate dispersion and cell dissocation or interference of cells in suspension with sedimenting soil particles. Clearly the former problem could be overcome by enhanced dissociation of cells and soil particles whereas the latter might be overcome by diluting the sample. The major advantage of differential centrifugation over elutriation is that the sample obtained is fairly concentrated (53).

Purification

Dense mineral materials in extracts from soils can be removed from samples using density-gradient centrifugation (5, 29, 69, 73). Osmotically balanced, coated colloidal silica media such as Percoll/sucrose gradients have been widely used in such procedures (44, 69, 73). The large difference in density between mineral materials and microorganisms means that separation can be obtained from a low organic matter soil using relatively rapid (<20 minutes at 10 000 g) rate-zonal centrifugation (53). Extracts from high organic matter soils cannot be so readily separated because of the lower density of organic matter (44, 53). However, not all of the cells present in extracts are completely dissociated from soil particles so significant numbers of cells are carried down the density gradient with the mineral particles (5, 53).

Populations of particles with differing variable charges can be separated using aqueous two-phase systems (3). Such an approach has been used for concentrating air-borne fungal spores and bacteria (14, 15, 116) and for separating *Penicillium chrysogenum* conidia from a suspension containing peat dust (117). Briefly, in this application a two-phase system consisting of a dextran-rich lower phase and an upper phase rich in polyethylene glycol and substituted charged sulphonylpolyethylene glycol was prepared. The system was buffered with phosphate buffer in the range of pH 1 to 5. The net charge and the hydrophobic properties of the cell surfaces seemed to be responsible for the partition behaviour (117). Partitioning of conidia from *Penicillium brevicompactum* and *Aspergillus fumigatus* was only moderately influenced by pH. For spores of *Rhizopus rhizopodiformis* and *Streptomyces griseus* and vegetative cells of *Bacillus subtilis*, increasing pH diminished the affinity for the upper phase (117). Careful manipulation of the pH led to the concentration of organic dust particles at the interface between the two phases. Such, an approach has not been evaluated for soils, but with a knowledge of the PZC of the variable charge components of soils and the surface characteristics of target microorganisms this may be a potentially useful approach to purifying soil extracts.

Co-ordinated sampling

Aspects of the dispersion, dissociation, separation and purification techniques discussed above have been assembled into the multi-stage extraction procedure by Hopkins *et al.* (53), which is shown in Fig. 5.1. After each dispersion/dissociation step the released microorganisms were separated by low-speed centrifugation and the resulting supernatants, when pooled,

SOIL SAMPLE

↓

MECHANICAL HOMOGENIZATION
(to disrupt large soil aggregates)

↓

SHAKING WITH ANIONIC DETERGENT
(to help overcome adhesion between organic
polymers, soil particles and microorganisms
AND CHELATING ION-EXCHANGE RESIN
(to disrupt soil particles)

↓

LOW-SPEED CENTRIFUGATION
(to remove released microorganisms)

↓

ULTRASONICATION
(to disrupt physical interactions between soil
particles and between soil particles and microorganisms)

↓

LOW-SPEED CENTRIFUGATION

↓

SHAKING WITH WATER
(with the intention of reducing
electrolyte concentration of the soil
suspension thereby increasing the repulsion
of like charged colloid surfaces)

↓

LOW-SPEED CENTRIFUGATION

↓

DENSITY-GRADIENT CENTRIFUGATION
(to separate released microorganisms
from less-dense soil particles)

Fig. 5.1 Schematic summary of the extraction and purification procedure used by Hopkins *et al.* (53). The diagram has been simplified by the omission of repeated steps.

constituted the extract. This extract was then purified by density-gradient centrifugation. The advantage of such a multi-stage extraction procedure was that it aimed to achieve maximum dispersion of soil aggregates and dissociation of microorganisms from soil particles by targeting a different aspect of the soil–microorganism interaction at each stage. The scanning electron micrographs of samples obtained from soil at different stage (Fig. 5.2) indicate the progressive nature of the sampling procedure. Such an approach may have helped restrict the damage to those microorganisms collected earlier in the extraction process. There was the added advantage that the rigour of each treatment could be increased progressively at each stage such that only those cells still retained in undispersed soil aggregates were subject to the potentially more damaging next step. For GEMs work, where it is desirable, that the extract contains the maximum number of organisms, this is probably desirable, since none of the cells left in the residues from earlier steps would be included in the final sample were it not for the increasing stringency of the extraction procedure.

Efficiency of extraction

There are two approaches to assessing the efficiency with which micro-organisms can be extracted from soil. It is important to know what proportion of the total soil microbial biomass has been sampled and whether the sample is representative of the indigenous microbial community. It is also important to know the proportion and nature of the non-biological soil material included in the sample. However, there are a variety of reasons why comparison between different sampling protocols is unsatisfactory. Quantitative recovery of microorganisms from soil has not been a major aim of many of the cited studies and in most cases reliable techniques for estimating the amount and physiological state of the microorganisms in extracts, residues and soil have not been available.

There are few reports on the final purity of soil bacterial extracts. Bakken (5) has shown that purity (in terms of the ratio of biomass N:total N) was greatest for a clay soil, least for an organic soil and intermediate for a sandy soil when extracts were prepared by homogenization and density-gradient centrifugation. This is probably because N in organo-mineral complexes in the clay soil was more readily separated by density gradient centrifugation than in the soils with low or zero clay contents. An alternative expression of purity, in which the proportion by weight of the dry matter in an extract was expressed relative to the dry weight of soil at the outset, has been used to show that the extract purity prior to density-gradient centrifugation was highest for a sandy soil (90%) and lowest for an organic soil (77%) (53). Further purification of the extracts by density-gradient centrifugation was limited because a large proportion of the cells were carried down the gradient with the mineral particles. Since purification of extracts can lead to losses in non-desorbed microorganisms (5,10,54) the concentration factor (CF) of the sample with microorganisms was used by Hopkins *et al.* (53) to assess purity. The CF was determined as the factor by which the ratio of microscopic cell count to soil material by weight was increased following extraction and purification. Values obtained were consistent with those of purity (53).

BIOMASS ESTIMATION

Many of the problems associated with estimates of recovery, extraction efficiency and extract purity are due to the difficulties in quantifying microbial biomass in natural habitats. Since it is impractical to provide isolation media and appropriate growth conditions for all microorganisms in a particular environment (21), viable counting procedures are unsuitable (18). Alternative approaches include direct microscopic counts and the use of specific chemical components of microbial cells. Examples of the recoveries of biomass for the procedure outlined in Fig. 5.1 are shown in Table 5.2 and discussed below.

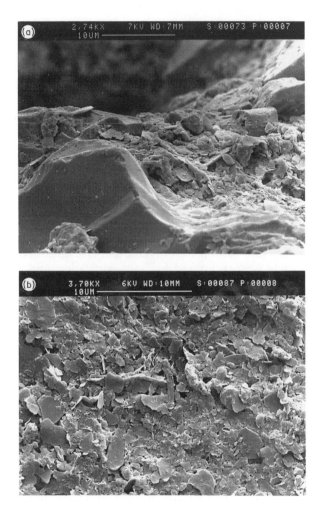

Fig. 5.2 Scanning electron micrographs of soil, soil extracts and residues at different stages of extraction. (See Fig. 5.1). (a), the surface of a soil aggregate at the outset; (b), soil extract prior to density-gradient centrifugation showing dispersed clay particles and organic debris but microorganisms not readily visible; (c), large soil particles retained in the residue; (d), bacteria and soil organic matter recovered as the buoyant fraction from density-gradient centifugation; (e), dispersed clay particles recovered as the dense fraction from density-gradient centrifugation.

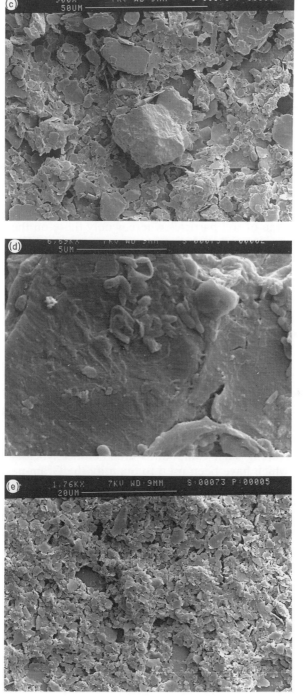

Table 5.2 Different estimates of percentage biomass recovery from soil (adopted from Ref. 53).

Soil	Direct microscopic count	ATP content	Phospholipid-P content	Lipopolysaccharide (β-hydroxymyristic acid) content	Ergosterol content	Viable (plate) count
Clay loam	61	26	41	24	0	28
Sandy loam	56	49	47	26	0	40
Peat	62	21	20	15	13	90

Accurate cell counts by fluorescence microscopy require that cells inside aggregates are visible (58, 94). This limitation leads to an inevitable over-estimate of recovery, since it is impossible to obtain an accurate figure for microbial numbers prior to extraction. Cells counted in the extracts are comparatively free of soil particles whereas only those on the outer surfaces of aggregates are counted in the intact soils and residues. The recovery may, therefore, be overestimated.

Biomass in soils, residues and extracts can be estimated by the extraction of specific chemical components such as ATP (118), and lipid-bound phosphate (30, 128). Such analyses can be considered relatively non-specific measures of biomass although lipid-bound phosphate can be regarded as a marker for prokaryotic biomass because of its relative abundance in such organisms. Furthermore, lipid-bound phosphates have a relatively high turnover rate in natural ecosystems as a result of which they are considered indicative of the active prokaryote biomass (34, 127). Although extraction efficiencies for lipid phosphates from sediments of 102% ± 20.5% have been recorded (30), preliminary studies in soils of different textures have suggested recoveries of less than 30% (53).

Other cell components can be used as more specific indicators of biomass composition. Among such compounds are: ketodeoxyoctonate and 3-OH-myristic acid, which have been used to quantify the Gram-negative biomass (91, 99); chitin (95, 126); ergosterol (126), used to measure fungal biomass; and fatty acids (76, 92). Although of obvious value in estimating the recovery of particular components of the soil biomass such procedures must be used cautiously since they are not always efficiently transferred between environments such as sediments and soils (126). For example, the soil microflora is numerically dominated by 'dwarf' cells (8) in which the relative amounts of most cellular components are reduced compared to those in normal sized cells (7). However, these cells maintain a full complement of DNA (6) and should be considered carefully when assessing biomass using cell structural components such as phospholipids.

For many of these analytical procedures the extraction efficiency has been measured by standard addition of the target compound. In assessing the efficacy of recovery procedures of indigenous microorganisms from soil, such standard addition is inappropriate. If added microorganisms are re-extracted immediately after addition, a greater proportion of the added than of the indigenous microorganisms are recovered in the extract (94). If time for the added microorganisms to penetrate soil aggregates, for internal rearrangements within aggregates to occur and for surface interactions to form is allowed between standard addition and re-extraction, there is scope for, on the one hand, growth of the added organisms, and on the other hand, death or predation of the added organisms. In either case standard addition experiments, in denying the complexity of soil–microbial interactions, should not be relied upon to provide an accurate estimate of extraction efficiency.

In monitoring studies, unless extensive steps are taken to ensure otherwise, a reduction in numbers of the target organism extracted from soil cannot be attributed solely to death of the organisms. It is possible that in studies running over a number of days or weeks introduced organisms may become more difficult to extract from the soil. The distinction between those microorganisms which are present but irrecoverable and those which have died needs to be considered in such studies.

TAXONOMIC ANALYSIS OF EXTRACTS

In recent years, our understanding of microbial processes and their importance and activity in soils has increased considerably due to the introduction of new techniques. However, information on the types of microorganisms which make up microbial communities remains limited. This is in part due to the fact that classical methods for studying microbial populations, in which the sample is shaken in diluent and then plated out on a 'non-selective' medium, are highly inefficient and, depending on the nature of the environment sampled, capable of recovering some 1% or less of the total cell numbers as judged using direct counts (18, 50).

To those interested in monitoring GEMs in natural environments it is important to be able to estimate the taxonomic composition of the microbial community into which the engineered organism will be introduced since this defines the genetic background against which gene transfer might occur (50). Such information also enables any qualitative or quantitative changes in species composition to be assessed.

Even if soil bacteria can be successfully extracted and cultured, there remains the problem of which procedures should be used to classify and identify the organisms. To ensure that information about a given habitat or the impact of an introduced organism on the environment is available on time it is important to be able to identify natural isolates quickly. Previous studies have shown, however, that limited phenetic analysis often results in the erroneous classification of bacterial isolates (22, 36), with most taxonomic surveys of microbial communities recovering the bacterial isolates in a relatively small number of major clusters (36). Numerical phenetic analyses on the other hand has been shown to be useful in delineating natural populations but suffers in that it can be time consuming. Furthermore, marker or reference strains have been shown to behave differently from recent isolates making cluster identification difficult (36). The identification of natural isolates therefore presupposes the existence of accurate classification and identification procedures. Such schemes are not available for all microbial groups and those which are await thorough testing with strains from a range of natural habitats.

Chemosystematics (chemotaxonomy or chemical taxonomy) is the study of the chemical variation in living organisms and the use of selected characters in classification and identification (37). It can be argued that all taxonomic techniques are in fact chemotaxonomic since morphology, serology, pigmentation and the biochemical properties of microorganisms are all influenced by the organisms' chemical composition. It is, however, useful to restrict the definition to the distribution of specific chemical components such as lipids, wall amino acids, sugars and proteins among microbial taxa. Such information can be used at all taxonomic levels although the discriminating power of particular components varies between taxa. The menaquinone composition of aerobic endospore-forming bacilli and the polar-lipid patterns of staphylococci are quite uniform and as such provide reliable generic markers. However, the menaquinone composition of the genus *Staphylococcus* and the polar-lipid patterns of the genus *Bacillus* show variation at the species level, making them valuable in sub-generic classification (80).

Many methods exist for the extraction and analysis of chemical components from microbial cells (37) and the speed and simplicity with which they can be applied make them potentially useful in the identification of environmental isolates. Most of the procedures used require little analytical chemical experience and can be performed with a minimum of equipment. A simple integrated extraction and analysis procedure for actinomycetes has been described previously (38, 85) and similar approaches have been used with diverse groups of Gram-positive and Gram-negative bacteria (37).

Chemical characters provide a simple means of identifying isolates to the generic level and are particularly useful when used in conjunction with other taxonomic techniques such as phenetic and genetic analysis. However, all such studies, whether based on biochemical test data, hybridization data or chemotaxonomic data, can only be used successfully if interpreted in the light of an existing classification. If such classifications do not exist, the data (i.e. the classification derived from the data) is by necessity habitat-specific.

Recently, sequence data from natural habitats have been proposed as a means of elucidating community structure (86, 104). In such studies, rRNA sequences are used to place previously unknown sequences within an existing phylogenetic framework (identification). The advantages of such an approach are that the individual components of the community do not need to be isolated prior to identification (104). However, the value of such sequence data, which can be costly and time-consuming to obtain, in routinely identifying environmental isolates to species level awaits testing, and for the present, species identification is best carried out by DNA-DNA hybridization or numerical taxonomy (22). Species identification, in the taxonomic sense

(56), is difficult to perform routinely but in many ecological studies it is the changes in species diversity which are important in understanding natural variability and its scales (33), a question central to the debate on the impact of GEMs on natural ecosystems.

Species identification using chemical methods

Chemical markers can be used to characterize isolates at the lower levels of taxonomic rank. Since this is a more difficult task the procedures required are generally more elaborate and labour intensive. Details of the type of chemotaxonomic methods for sub-generic and in some cases sub-specific identification of actinomycetes are listed in Table 5.3 (38). Unlike the analysis of key chemical components, these methods often require some analytical experience and access to suitable instrumentation.

Table 5.3 Chemical techniques used in the classification and identification of actinomycetes to specific and sub-specific levels. Nucleic acid hybridization techniques are not included (adopted from Ref. 38).

Anaylsis	Taxa	Statistics	Reference
Fatty acids	*Streptomyces*	SIMCA pattern recognition	98
Fatty acids	*Corynebacterium, Nocardia*	Cluster analysis	16
Fatty acids	Amycolate wall IV actinomycetes	Principal components	28
Whole cell proteins	*Corynebacterium*	Cluster analysis	55
Restriction patterns	*Mycobacterium*	None	93
Mycolic acid analysis	*Mycobacterium*	None	24

Fatty acids have been shown to be particularly useful in classification and identification when analysed using appropriate statistical techniques. Clustering of organisms according to their fatty acid composition has been carried out in several ways. In the majority of cases similarity values have been derived using a variety of coefficients (25, 78). These studies were based on diverse groups of organisms and addressed questions such as the importance of the correlation coefficient and the nature of the data transformation procedures (78). Most of these investigations have been concerned with the importance of fatty acids in the classification of microorganisms, but developments in the extraction and analysis of fatty acids (79, 84, 97, 98) have made it apparent that the reproducibility of the profiles and the ease and speed of analysis make such fatty acid data suitable for identification to the species level (16, 85) provided the conditions under which the organisms are cultured are carefully controlled. The latter has been shown to be particularly important with respect to those organisms which undergo major morphological changes such as sporulation or aerial mycelium formation during their growth cycle (97). Despite such limitations commercially available packages, comprising a gas-chromatograph, database and pattern recognition software are now available.

Pyrolysis techniques such as pyrolysis mass-spectrometry (103) and pyrolysis gas-chromatography (82) also have potential applications in the analysis of microbial communities. Pyrolysis is defined as the thermal degradation of complex organic molecules in an inert atmosphere to produce low-molecular weight fragments of the original sample. The fragments produced are characteristic of the original material and can be resolved to produce a fingerprint using either a gas-chromatograph or a mass-spectrometer. Pyrolysis data are complex and need to be analysed using multivariate statistical procedures (13, 70). Fortunately, such statistical routines are readily available with many of the packages suitable for personal computers (17).

The advantage of pyrolysis techniques is that they do not require any sample preparation, making it possible to fingerprint strains directly from primary isolation plates. When linked to a mass-spectrometer, each characterization can be performed in under 10 minutes, and previous studies have shown that depending on the statistical analysis, the data obtained can be used to differentiate to the species and even the sub-species level (83, 103).

The speed of identification, and the quality of the data produced using such an approach have obvious applications in microbial ecology. Recent developments in laser pyrolysis systems, in which the pyrolyser is used in conjunction with a microscope, makes possible the location, analysis and identification of microorganisms without the need for isolation from the environment. In this instrument, a laser is used to pyrolyse organisms viewed under the microscope. The pyrolysate is then conducted to a mass-spectrometer where it can be scanned and used to identify the single cell pyrolysed. At the moment, however, such instrumentation is expensive, largely experimental and not widely available.

CONCLUDING REMARKS

In the detection and analysis of specific taxonomic groups of microorganisms in environmental samples, it is important to distinguish between those cells which are present but irrecoverable and those that are absent. In this chapter we have tried to highlight the difficulties involved in extracting microorganisms from soil and in failing to appreciate the heterogeneity and spatial variability of soils of different textures. Direct probing techniques offer an alternative to extraction and recovery experiments but have not been extensively tested in complex environments such as soil. It is to be expected that the efficiency with which such *in situ* probe techniques can be used will also depend on the nature of the soil, particularly the relative amounts of clay and organic matter.

To date there are no routine, widely available classification and identification procedures capable of handling the diversity and numbers of microorganisms in complex environments such as soil. Many of the existing procedures require that the organisms are cultured prior to identification; identification implies the existence of a working classification. Molecular approaches do provide an alternative to classical microbial systematics and chemosystematics but the probe libraries and extensive databases necessary for identification are not yet widely available. It is also likely that in all but the simplest of habitats even the molecular approaches will prove too time-consuming for detailed microbial population analysis. Given these limitations it is necessary to question the importance of such detailed population studies in microbial ecology and to consider whether equally useful information can be obtained using procedures which indicate whether there has been a change in the indigenous populations over time or following the introduction of GEMs (33, 81). Such an approach is likely to be more sensitive than measuring changes in processes such as nitrification and nitrogen fixation and may provide an indication of the scale of variability caused by the introduction of GEMs.

ACKNOWLEDGEMENT

We are grateful to the UK Department of the Environment for financial support.

REFERENCES

1. Acea, M.J., Moore. C.R. & Alexander, M. (1988) *Soil Biology and Biochemistry* **20**, 509–515.
2. Ahmed, M. & Oades, J.M. (1984) *Soil Biology and Biochemistry* **16**, 465–470.
3. Albertsson, P.A. (1986) *Partition of Cells, Particles and Macromolecules*, 3rd edn. John Wiley & Sons, New York.
4. Archibald, A.R., Baddiley, J. & Hepinstall, S. (1973) *Biochimica et Biophysica Acta* **291**, 629–634.
5. Bakken, L.R. (1985) *Applied and Environmental Microbiology* **49**, 1482–1487.
6. Bakken, L.R. & Olsen, R.A. (1983) *Applied and Environmental Microbiology* **45**, 1188–1195.
7. Bakken, L.R. & Olsen, R.A. (1988) In *Perspectives in Microbial Ecology*, Eds Megusar, F. & Gantar, M. Slovene Society for Microbiology Ljubljana, pp. 561–566.
8. Balkwill, D.L. & Casida, L.E. (1973) *Journal of Bacteriology* **114**, 1319–1327.
9. Balkwill, D.L., Labeda, D.P. & Casida, L.E. (1975) *Canadian Journal of Microbiology* **21**, 252–262.
10. Balkwill, D.L., Rucinsky, T.E. & Casida, L.E. (1977) *Antonie van Leeuwenhoek* **43**, 73–87.
11. Bayer, M.E. & Sloyer, J.L. (1990) *Journal of General Microbiology* **136**, 867–874.
12. Beringer, J.E., Bennet, P.M. & Bale, M.J. (1989) In *Recent Advances in*

Microbial Ecology, Eds Hattori, T., Ishida, Y., Maruyama, Y., Morita, R. & Uchida, A. Japanese Scientific Societies Press, Tokyo, pp. 645–649.

13. Blomquist, G., Johansson, E., Soderstrom, B. & Wold, S (1979) *Journal of Analytical and Applied Pyrolysis* **1**, 53–65.
14. Blomquist, G.K., Palmgren, U. & Strom, G. (1984) *Scandinavian Journal of Work, Environment and Health* **10**, 253–258.
15. Blomquist, G.K., Strom, G.B. & Soderstrom, B. (1984) *Applied and Environmental Microbiology* **47**, 1316–1318.
16. Bousfield, I.J., Smith, G.L., Dando, T.R. & Hobbs, G. (1983) *Journal of General Microbiology* **129**, 375–394.
17. Bratchell, N., O'Donnell, A.G. & MacFie, H.J.H. (1989) In *Computers in Microbiology—A Practical Approach*, Eds Bryant, T. & Wimpenny, J. IRL Press, Oxford, pp. 41–63.
18. Brock, T.D. (1987) In *The Ecology of Microbial Communities*, Eds Fletcher, M.M., Gray, T.R.G. & Jones, J.G. Society for General Microbiology Symposium No. 41. Cambridge University Press, London, pp. 1–17.
19. Burns, R.G. (1977) In *Adhesion of Microorganisms to Surfaces*, Eds Ellwood, D.C. & Melling, J. Academic Press, London, pp. 109–138.
20. Cheshire, M.V., Sparling, G.P. & Mundie, C.M. (1983) *Journal of Soil Science* **34**, 105–112.
21. Colwell, R.R., Somerville, C., Knight, I. & Russell, A.D. (1988) In *The Release of Genetically Engineered Microorganisms*, Eds Sussman, M., Collins, C.H., Skinner, F.A., & Stewart-Tull, D.E. Academic Press, London, pp. 47–60.
22. Colwell, R.R., Steven, S., Hori, H., Muir, D., Conde, B., Straube, W. & Somerville, C. (1989) In *Recent Advances in Microbial Ecology*, Eds Hattori, T., Ishida, Y. Maruyama, Y., Morita, R.Y. & Uchida, A. Japanese Scientific Societies Press, Tokyo, pp. 653–657.
23. Derajagium, B.V. & Landau, L. (1941) *Acta Physiochem.* [URSS] **14**, 633–645.
24. Dobson, G., Minnikin, D.E., Minnikin, S.M., Parlett, J.H., Goodfellow, M., Ridell, M. & Magnusson, M. (1985) In *Chemical Methods in Bacterial Systematics*, Eds Goodfellow, M. & Minnikin, D.E. Academic Press, London, pp. 238–265.
25. Drucker, D.B. (1974) *Canadian Journal of Microbiology* **20**, 1723–1728.
26. Edwards, A.P. & Bremner, J.M. (1965) *Nature* **205**, 208–209.
27. El-Falaha, B.M.A., Furr, J.R. & Russell, A.D. (1989) *Letters in Applied Microbiology* **8**, 15–19.
28. Embley T.M., Smida, J. & Stackebrandt, E. (1988) *Journal of General Microbiology* **134**, 961–966.
29. Faegri, A., Torsvik, V.L. & Goskøyr, J. (1977) *Soil Biology and Biochemistry* **9**, 105–112.
30. Findlay, R.H., King, G.M. & Whatling, L. (1989) *Applied and Environmental Microbiology* **55**, 2888–2893.
31. Ford, S. & Olson, B.H. (1988) In *Advances in Microbial Ecology* **10**, Ed. Marshall, K.C. Plenum Press, New York and London, pp. 45–79.
32. Foster, R.C. (1986) *Biology and Fertility of Soils* **6**, 189–203.
33. Fuhrman, J.A. & Lee, S.H. (1989) In *Recent Advances in Microbial Ecology*, Eds Hattori, T., Ishida, Y., Maruyama, Y., Morita, R.Y. & Uchida, A. Japanese Scientific Societies Press, Tokyo, pp. 687–691.
34. Gehron M.J. & White, D.C. (1983) *Journal of Microbiological Methods* **1**, 23–32.

35. Glaser, D., Keith, T., Riley, P., Chambers, A., Manning, J., Hattingh, S. & Evan, R. (1986) In *Monitoring Techniques for Genetically Engineered Microorganisms in Biotechnology and the Environment, Research Needs*, Eds Omenn, A.S. & Teich, A.H. Noyes Data Corporation, New Jersey.
36. Goodfellow, M. & Dickinson, C.H. (1985) In *Computer Assisted Bacterial Systematics*, Eds Goodfellow, M., Jones, D. & Priest, F.G. Academic Press, London, pp. 165–225.
37. Goodfellow, M. & Minnikin, D.E. (1985) *Chemical Methods in Bacterial Systematics*, Academic Press, London.
38. Goodfellow, M. & O'Donnell, A.G. (1989) In *Microbial Products: New Approaches*, Eds Baumberg, S., Hunter, I. & Rhodes, M. Society for General Microbiology Symposium No, 44. Cambridge University Press, Cambridge. pp. 343–383.
39. Graham, J.M. (1978) In *Centrifugal Separations in Molecular and Cell Biology* Eds Birnie, G.D. & Rickwood, D. Butterworths, London, pp. 63–114.
40. Gray, T.R.G. Baxby, P., Hill, T.R. & Goodfellow, M. (1968) In *The Ecology of Soil Bacteria*, Eds Gray, T.R.G. & Parkinson, D. Liverpool University Press, Liverpool, pp. 171–192.
41. Gray, T.R.G.,(1976) In *The Survival of Vegetative Microbes*, Eds Gray, T.R.G. & Postgate, J.R. Society for General Microbiology Symposium No. 26, Cambridge University Press, Cambridge, pp. 327–364.
42. Gregorich, E.G., Kachanoski, R.G. & Voroney, R.P. (1988) *Canadian Journal of Soil Science* **68**, 395–403.
43. Griffin, D.M. (1981) In *Water Potential Relationships in Soil Microbiology*, Eds Elliot, L.F., Papendick, R.I. & Wildung, R.E. Soil Science Society of America Special Publication No. 9. SSSA Madison, Wisconsin, pp. 141–151.
44. Griffiths, B.S. & Ritz, K. (1988) *Soil Biology and Biochemistry* **20**, 163–173.
45. Grim, R.E. (1968) *Clay Mineralogy* 2nd edn. McGraw Hill, New York.
46. Hattori, T. (1988) *Report of the Institute of Agricultural Research, Tohoku University.*, **37**, 23–36.
47. Hattori, T. & Hattori, R. (1976) CRC *Critical Reviews in Microbiology* **4**, 423–461.
48. Herron, P.R. & Wellington, E.M.H. (1990) *Applied and Environmental Microbiology* **56**, 1406–1412.
49. Heynen, C.E., van Elsas, J.D., Kuikman, P.J. & van Veen, J.A. (1988) *Soil Biology and Biochemistry* **20**, 483–488.
50. Hofle, M.G. (1989) In *Recent Advances in Microbial Ecology*, Eds Hattori, T., Ishida, Y., Maruyama, Y., Morita, R.Y. & Uchida, A. Japanese Scientific Societies Press, Tokyo, pp. 692–696.
51. Holben, W.E., Jansson, J.K., Chelm, B.K. & Tiedje, J.M. (1988) *Applied and Environmental Microbiology* **54**, 703–711.
52. Hopkins, D.W., O'Donnell, A.G. & Shiel R.S. (1989) *Biology and Fertility of Soils* **8**, 335–338.
53. Hopkins, D.W., Macnaughton, S.J. & O'Donnell, A.G. (1991) *Soil Biology and Biochemistry* **23**, 217–225.
54. Hopkins, D.W., O'Donnell, A.G. & Macnaughton, S.J. (1991) *Soil Biology and Biochemistry* **23**, 227–232.
55. Jackman, P.J.H. (1982) *Journal of Medical Microbiology* **15**, 485–492.
56. Jones, D. & Sackin, M.J. (1980) In *Microbiological Classification and Identification*, Eds Goodfellow, M. & Board, R.G. Academic Press, London, pp. 73–106.

57. Jeng, D.K., Lin, L.I. & Hervey, L.V. (1990) *Journal of Applied Bacteriology* **68**, 479–484.
58. Jenkinson, D.S., Powlson, D.S. & Wedderburn, R.W.M. (1976) *Soil Biology and Biochemistry* **8**, 189–202.
59. Kanazawa, S., Takeshima, S. & Ohta, K. (1986) *Soil Science and Plant Nutrition* **32**, 81–89.
60. Keya, S.O. & Alexander, M. (1975) *Soil Biology and Biochemistry* **7**, 231–237.
61. Kilbertus, G. (1980) *Revue d'Ecologie et Biologie du Sol* **17**, 543–557.
62. Kilbertus, G., Proth, J. & Verdier, B. (1979) *Soil Biology and Biochemistry* **11**, 109–114.
63. Ladd, J.N. (1989) In *Recent Advances in Microbial Ecology*, Eds Hattori, T., Ishida, Y., Maruyama, Y., Morita, R.Y. & Uchida, A. Japanese Scientific Societies Press, Tokyo, pp. 169–174.
64. Lahav, N. (1962) *Plant and Soil* **17**, 191–208.
65. LaMondia, J.A. & Brodie, B.B. (1987) *Journal of Nematology* **19**, 104–107.
66. van Loosdrecht, M.C.M., Lyklema, J., Norde, W. & Zehnder, A.J.B. (1989) *Microbial Ecology* **17**, 1–15.
67. Lynch, J.M. (1981) *Journal of General Microbiology* **126**, 371–375.
68. Macdonald, R.M. (1986) *Soil Biology and Biochemistry* **18**, 399–406.
69. Macdonald, R.M. (1986) *Soil Biology and Biochemistry* **18**, 407–410.
70. MacFie, H.J.H. & Gutteridge, C.S. (1982) *Journal of Analytical and Applied Pyrolysis* **4**, 175–204.
71. Marshall, K.C. (1967) *Australian Journal of Biological Sciences* **20**, 429–435.
72. Marshall, K.C. (1976) *Interfaces in Microbial Ecology*. Harvard University Press, Cambridge, Mass., and London, UK.
73. Martin, N.J. & Macdonald, R.M. (1981) *Journal of Applied Bacteriology* **51**, 243–251.
74. McNown, J.S. & Malaika, J. (1950) *Transactions of the American Geophysical Union* **31**, 74–78.
75. Molope, M.M., Grieve, I.C. & Page, E.R. (1987) *Journal of Soil Science* **38**, 71–78.
76. Nichols, P.D., Smith, G.A., Antworth, C.P., Hanson, R.S. & White, D.C. (1985) *FEMS Microbiology Ecology* **31**, 327–335.
77. Niepold, F., Conrad, R. & Schlegel, H.G. (1979) *Antonie van Leeuwenhoek* **45**, 485–497.
78. O'Donnell, A.G. (1985) In *Computer Assisted Bacterial Systematics*, Eds Goodfellow, M., Jones, D. & Priest, F.G. Academic Press, London, pp. 403–414.
79. O'Donnell, A.G. (1986) In *Biological, Biochemical and Biomedical Aspects of Actinomycetes*, Eds Szabo, G., Biro, S. & Goodfellow, M. Akademai Kiado, Budapest, pp. 541–549.
80. O'Donnell, A.G. (1988) In *Acinomycetes in Biotechnology*, Eds Goodfellow, M., Williams, S.T. & Mordarski, M. Academic Press, London, pp. 69–88.
81. O'Donnell, A.G. (1989) In *Recent Advances in Microbial Ecology*, Eds Hattori, T., Ishida, Y., Maruyama, Y., Morita, R.Y. & Uchida, A. Japanese Scientific Societies Press, Tokyo, pp. 674–678.
82. O'Donnell, A.G. & Norris, J.R. (1981) In *The Aerobic Endospore-Forming Bacteria*, Eds Berkeley, R.C.W. & Goodfellow, M. Academic Press, London, pp. 141–179.
83. O'Donnell, A.G., Norris, J.R., Berkeley, R.C.W., Claus, D., Kaneko, T.,

Logan, N.A. & Nozaki, R. (1980) *International Journal of Systematic Bacteriology* **30**, 448–459.

84. O'Donnell, A.G., Minnikin, D.E. & Goodfellow, M. (1985) In *Chemical Methods in Bacterial Systematics*, Eds Goodfellow M. & Minnikin, D.E. Academic Press, London, pp. 131–143.

85. O'Donnell, A.G., Nahaie, M.R., Goodfellow, M., Minnikin, D.E. & Hajek, V. (1985) *Journal of General Microbiology* **131**, 2023–2033.

86. Olsen, G.J., Lane, D.J., Giovannoni, S.J., Pace, N.R. & Stahl, D.A. (1986) *Annual Review of Microbiology* **40**, 337–365.

87. Oostenbrink, M. (1960) In *Nematology* Eds Sasser, J.N. & Jenkins, W.R. University of North Carolina Press, Chapel Hill, pp. 85–162.

88. Page, A.L., Miller, R.H. & Keeney, D.R. (1982) *Methods of Soil Analysis, Part 2: Chemical and Microbiological Properties* Agronomy Monograph No. 9. American Society of Agronomy, Soil Science Society of America, Crop Science Society of America, Madison, Wisconsin.

89. Papendick, R.I. & Campbell, G.S. (1981) In *Water Potential Relationships in Soil Microbiology*, Eds Elliot, L.F., Papendick, R.I. & Wildung, R.E. Soil Science Society of America Special Publication No. 9. SSSA Madison, Wisconsin, pp. 1–22.

90. Parfitt, R.L. (1980) In *Soils with Variable Charge*, Ed. Theng, B.K.G. New Zealand Society of Soil Science, Lower Hutt, pp. 167–194.

91. Parker, J.H., Smith, G.A., Fredrickson, H.L., Vestal, J.R. & White, D.C. (1982) *Applied and Environmental Microbiology* **44**, 1170–1177.

92. Parkes, R.J. (1987) In *The Ecology of Microbial Communities*, Eds Fletcher, M.M., Gray, T.R.G. & Jones, J.G. Society for General Microbiology Symposium No. 41. Cambridge University Press, London, pp. 147–177.

93. Patel, R., Kvach, J.T. & Mounts, P. (1986) *Journal of General Microbiology* **132**, 541–551.

94. Ramsey, A.J. (1984) *Soil Biology and Biochemistry* **16**, 475–481.

95. Ride, J.P. & Drysdale, R.B. (1972) *Physiology and Plant Pathology* **2**, 7–15.

96. Rutter, P.R. & Vincent, B. (1980) In *Microbial Adhesion to Surfaces*, Eds Berkeley, R.C.W., Lynch, J.M., Melling, J., Rutter, P.R. & Vincent, B. Ellis Horwood, Chichester, pp. 79–92.

97. Saddler, G.S., O'Donnell, A.G. & Goodfellow, M. (1985) *Journal of Applied Bacteriology* **60**, 51–56.

98. Saddler, G.S., O'Donnell, A.G., Goodfellow, M. & Minnikin, D.E. (1987) *Journal of General Microbiology* **133**, 1137–1147.

99. Saddler, J.N. & Wardlaw, A.C. (1980) *Antonie van Leeuwenhoek* **46**, 27–39.

100. Scheraga, M., Meskill, M & Litchfield, C.D. (1979) In *Methodology of Biomass Determinations and Microbial Activities in Sediments*, Eds Litchfield, D. & Seyfriend, P.L. American Society for Testing and Materials, pp. 21–39.

101. Schuttler, P.L. & Weaver, T. (1986) *Soil Biology and Biochemistry* **18**, 389–394.

102. Seinhorst, J.W. (1962) *Nematologia* **8**, 117–128.

103. Schute, L.A., Berkeley, R.C.W., Norris, J.R. & Gutteridge, C.S. (1985) In *Chemical Methods In Bacterial Systematics*, Eds Goodfellow, M. & Minnikin, D.E. Academic Press, London, pp. 95–114.

104. Stahl, D.A., Flesher, B., Mansfield, H.R. & Montgomery, L. (1988) *Applied and Environmental Microbiology* **54**, 1079–1084.

105. Stahl, D.A., Devereux, R., Amann, R.I., Flesher, B., Lin, C. & Stromley, J.

(1989) In *Recent Advances in Microbial Ecology*, Eds Hattori, T., Ishida, Y., Maruyama, Y, Morita, R. & Uchida, A. Japanese Scientific Societies Press, Tokyo, pp. 669–673.

106. Steffan, R.J., Goksøyr, J., Bej, A.K. & Atlas, R.M. (1988) *Applied and Environmental Microbiology* **54**, 2908–2915.
107. Stevenson, F.J. (1982) *Humus Chemistry*, Academic Press, London.
108. Stotzky, G. (1966) *Canadian Journal of Microbiology* **12**, 831–848.
109. Stotzky, G. (1966) *Canadian Journal of Microbiology* **12**, 1235–1246.
110. Stotzky, G. & Rem, L.T. (1966) *Canadian Journal of Microbiology* **12**, 547–563.
111. Stotzky, G. (1980) In *Microbial Adhesion to Surfaces*, Eds Berkeley, R.C.W., Lynch, J.M., Melling, J., Rutter, P.R. & Vincent, B. Ellis Horwood, Chichester, pp. 231–247.
112. Stotzky, G. (1986) In *Interactions of Soil Minerals with Natural Organics and Microbes*, Eds Huang, P.M. & Schnitzer, M. Soil Science Society of America Publication No. 17. Soil Science Society of America, Madison, Wisconsin, pp. 305–428.
113. Stotzky, G. & Babich, H. (1986) *Advances in Applied Microbiology* **31**, 93–138.
114. Stotzky, G. & Burns, R.G. (1982) In *Experimental Microbiology Ecology*, Eds Burns, R.G. & Slater, J.H. Blackwell Scientific Publications, Oxford, pp. 105–133.
115. Strickland, T.C., Sollins, P., Shimel, D.S. & Kearle, E.A. (1988) *Soil Science Society of America Journal* **52**, 829–833.
116. Strom, G.B. & Blomquist, G.K. (1986) *Applied and Environmental Microbiology* **52**, 723–726.
117. Strom, G.B., Palmgren, U. & Blomquist, G. (1987) *Applied and Environmental Microbiology* **53**, 860–863.
118. Tate, K.R. & Jenkinson, D.S. (1982) *Soil Biology and Biochemistry* **14**, 331–335.
119. Theng, B.K.G. (1979) *Formation and Properties of Clay-Polymer Complexes*. Elsevier Scientific Publishing, Amsterdam.
120. Thompson, M.L., Zhang, H., Kazemi, M. & Sandor, J.A. (1989) *Soil Science* **148**, 250–257.
121. Tommerup, I.C. (1982) *Applied and Environmental Microbiology* **41**, 533–539.
122. Tommerup, I.C. & Carter, D.J. (1982) *Soil Biology and Biochemistry* **14**, 69–71.
123. van Veen, J.A. & van Elsas, J.D. (1988) In *Perspectives in Microbial Ecology*, Eds Megusar, F. & Gantar, M. Slovene Society for Microbiology, Ljubljana, pp. 481–488.
124. Verwey, E.J.W. & Overbeek, J.T.G. (1948) *Theory of the Stability of Lyophobic Colloids*. Elsevier, Amsterdam.
125. Waites, W.M. & Bayliss, C.E. (1984) In *Revival of Injured Microbes*, Eds Andrew, M.H.E. & Russell, A.E. Academic Press, London, pp. 221–240.
126. West, A.W. & Grant, W.D. (1986) *Journal of Microbiological Methods* **6**, 47–53.
127. White, D.C., Bobbie, R.J., Morrison, S.J., Oosterhof, D.K., Taylor, C.J. & Meeter, D.A. (1977) *Oceanography* **22**, 1089–1099.
128. White, D.C., Davis, W.M., Nickels, J.S., King, J.D. & Bobbie R.J. (1979) *Oecologia* **40**, 51–62.
129. Williams, S.T., Parkinson, D & Burges, N.A. (1965) *Plant and Soil* **22**, 163–186.

Chapter 6

Gene Transfer in Terrestrial Environments and the Survival of Bacterial Inoculants in Soil

E.M.H. Wellington, P.R. Herron and N. Cresswell
Department of Biological Sciences, University of Warwick

INTRODUCTION

Gene transfer in nature has been investigated using two distinct approaches. The first involved the detection of homologous genes in distantly related bacteria where horizontal transfer could account for such anomalies, particularly if the genes were found on plasmids or transposons (for recent review see Ref. 169). The second approach has been to demonstrate gene transfer experimentally under natural conditions, or in model systems which attempt to simulate some of the conditions found in the natural habitat. The second approach is considered in this chapter, which reviews the experimental methods and data collected in the study of genetic interactions between bacteria in soil, and the fate of genetically engineered bacterial inoculants and introduced genes.

Monitoring Genetically Manipulated Microorganisms in the Environment. Edited by C. Edwards
Published 1993 John Wiley & Sons Ltd. © 1993 E.M.H. Wellington, P.R. Herron and N. Cresswell

Soils frequently contain large numbers of microorganisms, many of which have not been isolated or characterized. Bacterial and fungal spores can reside in the soil for long periods of time (110); other cells may also be inactive but differ only slightly in morphology from their active counterparts. Soil is not a homogeneous substrate but extremely variable in its physical and chemical structure, with biological material such as plants and animals being an integral part of the environment. A consequence of such heterogeneity is that variations at the level of the microenvironment will be significant and provide a wide range of niches for colonization. Predicting the fate of a genetically engineered microorganism (GEM) in soil will be a difficult task given that we are dealing with a largely uncharacterized population of microorganisms in a highly variable habitat. Stotzky (156) listed a wide range of environmental factors likely to be important in determining the fate of a bacterial inoculant in soil. Numerous studies with soil microcosms indicate that nutrient availability is one of the most important factors affecting the frequency and occurrence of gene transfer.

Bulk soil is a nutrient-poor environment but with areas of enhanced nutrient status. Van Elsas (169) emphasized the role of nutrient availability in determining the extent of gene transfer and noted that transfer was most likely to occur in sites such as the rhizosphere, on or in plants or soil animals and on degrading organic matter. The supply of nutrients to the soil environment is sporadic and varied, which makes it difficult to emulate in laboratory-based studies.

With the exception of photosynthetic microorganisms at the soil surface and other autotrophs within the soil bulk most of the soil microflora is heterotrophic. In soil, the growth-limiting substrate is generally thought to be nitrogen. The supply of nitrogen may be from fresh leaf litter, root exudates or dead and dying organisms. Bacteria are also an important source of organic nitrogen; they have many predators (190) and also allow cryptic growth of other bacteria (28). The arrival of plant litter at the soil surface, therefore, provides microbes with a variety of energy sources, some of which are readily available while others require extensive degradation (185). The second source of nutrients for soil microorganisms is from the death of plant roots, the sloughing off of root cells as the root moves through the soil and root exudation of organic molecules through the root cell wall. Initially, nutrients are decomposed at the soil or root surface before being transferred into the soil body by leaching, diffusion or by soil animals (3). Thus, these substrates are present in discrete packages by gradients of soluble nutrients for a limited period of time prior to their utilization by the soil microflora (112, 185). This introduces temporal as well as spatial heterogeneity into the system. Whilst some microorganisms utilize soluble substrates in the soil water layer, others make use of the discrete bundles of organic matter. The survival of a bacterium in its natural habitat depends on its ability to grow at a rate sufficient to balance that of death from starvation and other natural causes (39). Soil may be regarded as an oligotrophic environment, in that the overall level of nutrients is low (185). However, the situation is complicated by the spatial and temporal variation in nutrient levels. In aquatic systems there is less chance of nutrient pockets and so many of the indigenous bacteria are oligotrophic (122). There is little unequivocal evidence for the existence of obligate oligotrophs in the terrestrial environment (185).

The fate of a GEM or an introduced gene will therefore depend on whether the inoculant can survive, grow and subsequently interact with the indigenous bacterial community. The fate of the gene in the indigenous community is dependent upon its selective value, extent of expression and environmental selection pressures. Stotzky (156) emphasized the discreteness of microhabitats in soil, which provides a further limitation to gene transfer. Cell-to-cell contact will be reduced if cells are physically isolated; filamentous microorganisms are able to bridge pore spaces and so may be more likely to interact genetically.

STUDYING THE FATE OF INTRODUCED GENES IN SOIL

Field studies

Field soils show extensive variability and complexity and are subject to uncontrolled climatic changes. This has made the use of model soil systems essential for studying the effects of individual environmental factors on gene transfer under controlled conditions. There are also restrictions involved in the release of GEMs into the environment. As a result experimental research into the fate of inoculants and their genes has been largely restricted to the use of model systems. Only a few field experiments and trials have been done in the natural environment. Of these the best documented in the UK was the testing of the modified baculovirus as an insecticide used for the control of pine beauty moth. The development and evaluation of this engineered virus have been well documented (13). Field studies have also been conducted by Hirsch & Spokes (76), following the fate of a Tn5-marked *Rhizobium* strain. Other major field studies were performed in the USA and Europe. One of the first in the USA involved in the Ice⁻ strain of *Pseudomonas syringae* developed by Lindow (95) and colleagues. The field trial involved the controlled release of a marked non-ice-nucleating strain developed to reduce frost damage on crops by competitive exclusion of ice-nucleating phylloplane bacteria. Other field studies were designed to demonstrate the use of markers such as *lacZY* for the efficient detection and monitoring of survival and spread (44). To date no field experiments have reported any cases of gene transfer to indigenous populations.

Model microcosms

Model systems have been used extensively to study soil microbial processes such as xenobiotic decompositions, nitrogen transformation, soil respiration and nutrient leaching. Gould *et al.* (60) designed sterile soil microcosms to study microbial breakdown of chitin in soil, and Blair *et al.* (14) investigated the effect of naphthalene on microbial respiration and nitrogen pools in soil-litter microcosms. The latter system was designed to simulate leaf litter standing stocks of 300 g m^{-2}. Burns (21) reviewed the use of experimental models for studying soil microbiology and outlined two basic types related to the use of intact or homogenized soil. Both types can be run as closed or open systems, the latter are useful for continuous measurements (see, for example, Ref. 4). Most of the early investigations into gene transfer were conducted using sterile soil microcosms (61, 62, 177). Once the soil has been sterilized, by whatever means, it becomes artificially amended, with increased levels of organic carbon and nitrogen (126). Sterilization of the soil allows introduced inoculants to colonize and grow in the system without competition and interactions with the natural microbial population. As a result these studies can only be used to indicate that genetic interactions can take place between given inocula in a model soil system.

The size of microcosms used has varied from 2 g (41, 165, 191), 3 g (65) to around 12 kg, used in non-sterile systems by Henschke *et al.* (72) during survival studies of a strain of *Pseudomonas fluorescens* harbouring a plasmid. Microcosms for the study of survival and gene transfer between bacteria have also been designed to include plants and invertebrates (12, 52, 71, 72, 170).

Homogenized soil

The simplest microcosm designs used air-dried sieved soil wetted to a selected matric potential or water content. Mesh size used varied between 1.65 mm (15), 2 mm (31, 94, 182) and 4 mm (132, 166, 179). Most of the microcosms used for the study of gene transfer were closed systems with aliquots of dried, sieved soil placed in screw-cap test-tubes (88), bottles (74), glass columns (73) and plastic boxes (182). Henschke & Schmidt (71) used a closed column containing a plant to study the survival of bacteria and their plasmids in soil. Open systems have been used to study transport of *Pseudomonas fluorescens* through soil columns (166) and have allowed detection of plasmid transfer to an inoculant pseudomonad population (19). In the latter study the percolation water was collected and sampled. Watering the soil allowed addition of 2,2-dichloropropionate; this gave a selective advantage to cells harbouring a plasmid coding for the breakdown of this substrate. Distinct rounds of replication in response to pulse addition of nutrients was achieved with a fed-batch microcosm described by Cresswell *et al.* (35) for monitoring gene transfer between streptomycetes in soil (see Figs 6.2 and 6.3). Systems which allow repeated rounds of cell division and replication may be useful for investigations of selection in soil and the subsequent amplification of a rare gene transfer event.

Intact undisturbed soil

Microcosms containing unsieved soil have been designed primarily for the study of transport and survival of GEMs in soil. Bentjen *et al.* (12) prepared intact soil-core microcosms (17.5 cm × 60 cm) using a steel coring apparatus; the same microcosm design was used by Fredrickson *et al.* (52). These microcosms were also designed to include plants, insects and invertebrates. The cores were used to study the fate of *Azospirillum lipoferum* and *Pseudomonas* sp. Tn*5* mutants in soil. Components sampled for the presence of the inocula included microcosm leachate, bulk soil, rhizosphere soil, rhizoplane, earthworm gut, insects and xylem exudate. The influence of plant rhizosphere on the survival of, and conjugation between, pseudomonads has also been studied (172) using a soil chamber first described by Dijkstra *et al.* (43). This microcosm design involved the separation of wheat seedling roots from the soil using a nylon gauze, so that the soil beneath the root mat could be sampled at various depths from the roots using a thin slicing technique. A detailed study by Bolton *et al.* (16, 17) used intact soil-core microcosms to compare with field lysimeters and field plots to provide a field calibration for their microcosms. Data on the survival, distribution and activity of a genetically altered pseudomonad were collected for the field and soil-core systems, in addition to analysis of soil ecosystem structure and function (17). These studies revealed that the soil-core microcosms were reliable indicators of field results when incubated at average field temperatures but were not so realistic after continuous incubation at 22°C.

Soil amendment

Clay

Clay amendment for the study of gene transfer and inoculant survival was first used extensively in studies by Weinberg & Stotzky (177). Six different soils were used; two were amended with 10% and 20% montmorillonite. Conjugation experiments using these soils showed that amendment enhanced the relative frequencies of conjugation compared to those observed in natural soil. The selection of clay amendment depends on the type of soil used in the microcosm systems. Van Elsas *et al.* (171) amended 10 g soil microcosms with bentonite clay. Amendment with bentonite and nutrients (tryptone yeast extract broth) gave significantly higher transfer frequencies for crosses with *Bacillus subtilis* and *B. cereus* in sterile soil when compared with those determined in soil amended with nutrients only. In non-sterile soil transfer was most observed without the addition of bentonite and nutrients. Van Elsas *et al.* (173) also used bentonite amendment in soil microcosms, which enhanced inoculant survival. Richaume *et al.* (136) demonstrated that montmorillonite amendment (15% w/w) increased plasmid transfer between *Escherichia coli* and *Rhizobium freudii*. Clay amendment has also been used in the study of *E. coli* phage infection in soil; Zeph *et al.* (191) amended soil with montmorillonite and kaolinite by up to 12% (vol/vol) but found that, unlike conjugation, transduction was not significantly enhanced. In addition, experiments were carried out with nutrient additions (modified Luria broth containing 0.1% glucose), which also failed to increase transduction frequencies.

Nutrients

The types of nutrient amendment used in soil microcosm systems, whether sterile or non-sterile, fall into two broad groups. The first comprises laboratory growth media such as nutrient broth, tryptone yeast extract broth and L-broth. These types of amendments provide the inoculants with high levels of readily available carbon and nitrogen sources. Trevors & Oddie (163) did not detect R-plasmid transfer between *E. coli* in sterile unamended soil; transfer was only detected when sterile soil (2 g) was amended with nutrient broth. The growth of inoculants under such conditions in sterile soil is far removed from the natural environment.

The selective effects of nutrients in soil on transfer frequencies of plasmids bearing heavy metal resistance genes was investigated by Top *et al*. (159). In plate matings a non-conjugative, mobilizable plasmid, pDN705p, bearing the *czc* genes was mobilized by conjugative plasmids RP4 or pULB113. Plasmid-borne resistance to cadmium, cobalt and zinc (*czc*) was transferred by conjugation in soil from *E. coli* to *Alcaligenes eutrophus*. Both strains were inoculated at quite high levels (10^6 g^{-1} soil) but selection of resistant transconjugants was achieved as *E. coli* did not express the *czc* genes. Transfer frequencies in homogenized non-sterile, nutrient-amended soil were around 10^{-7} but increased to 10^{-5} in sterile soil and were enhanced by nutrient amendment. Mobilization of pDN705 was dependent on the soil type and was directly related to available nutrient levels.

The second approach to nutrient amendment has been the use of more natural substrates with lower levels of readily available C and N. Van Elsas *et al*. (172) amended soil chamber microcosms with 0.1% and 1.0% freshly ground wheat plant tissue. Richaume *et al*. (136) found that amendment with ground corn stover enhanced plasmid transfer and an amendment at 5% (w/w) was optimal. The type of nutrient addition depends on the inoculants being studied. *Streptomyces* can utilize a large array of complex carbon sources, which are often used to selectively isolate members of this genus from soil; for example, chitin medium (6, 78) and starch casein medium (6,186). Carbon sources such as chitin (15, 178, 179), starch (178, 179), cottonseed flour and cellulose (15) have been used to nutrient amend soil microcosms in the study of growth, survival and plasmid transfer between streptomycetes in soil. An advantage with actinomycetes and other indigenous soil bacteria is that many of these bacteria grow well on semi-degraded plant materials and complex C and N sources. Nutrient amendments with unnaturally high levels of sugars and amino acids may distort the indigenous soil population and give advantage to faster growing bacteria such as enteric groups not normally present in significant numbers in natural soils.

Moisture, temperature and pH

In addition to nutrient status other physical and chemical soil attributes have been varied. In studies of genetic interactions between bacteria in soil the moisture, temperature and pH status have received some attention although there are many other variables such as density, porosity, and mineral and humus content.

Soil moisture

Water availability is an important factor when studying bacterial survival. Soil water content can be expressed in different ways; some, for example percent w/w, are difficult to interpret as different soils have different moisture-holding capacities. It is important to know the actual amount of available water, which is best expressed by water tension in kPa or bar.

In most studies a soil moisture content most suitable for the inoculants under test has been selected. A wide range of studies have used 60% moisture-holding capacity (MHC), which involved the addition of variable amounts of water depending on the ability of the soil to absorb water. For studies with non-indigenous, more drought-sensitive bacteria, such as *E. coli*, soil moisture tensions of −33 kPa have been used (88). This matric potential would be equivalent to around 20% (w/w) water content for a loam soil, but nearer to 50% (w/w) for a clay soil.

Studies with drought-resistant bacteria have been carried out at lower moisture tensions; for example, streptomycetes will survive and grow in low-moisture environments. Wellington *et al.* (178, 179) detected gene transfer between streptomycete inocula in sterile and non-sterile soil with a moisture content of 40% MHC at −67 kPA. Bleakley & Crawford (15) also studied conjugation between streptomycetes in soil microcosms over moisture contents in the range 20–60% MHC. Results indicated that with nutrient amendment the highest transfer frequencies were obtained at 20% MHC for a silt loam soil. It is not possible to compare this directly with studies using matric potential but the conditions would probably be very dry (perhaps −600 kPa).

Incubation temperature

Soil microcosms are frequently incubated at temperatures nearer to the optimal for growth of exotic inoculants rather than at temperatures representative of average soil conditions in the field. The field studies of Bolton *et al.* (16, 17) have clearly demonstrated the importance of using average field soil temperatures. No significant transfer of plasmids has been detected below 10°C using either *E. coli* as donor and recipient of R-plasmids (164) or in studies using *Pseudomonas* strains as donors and recipients of RP4 (165). However, Richaume *et al.* (136) found some transfer occurred between *E. coli* and *R. fredii* at 5°C (transfer frequency at 6.5×10^{-7}) with high levels of donor and recipient ($\sim 10^8$–10^9 cfu per g dry sterile soil).

Optimal temperatures for gene transfer in soil microcosms have proved to be between 20 and 30°C. Gene transfer was detected at both 37°C (61) and 40°C (136) for bacilli and enteric inoculants.

Soil pH

The pH of soil microcosms used for studying gene transfer has been adjusted using CaCO$_3$ (15), HCl or NaOH (164). Most studies indicated that a soil pH close to neutrality was optimal for gene transfer, and this was also the optimum pH for growth of the inoculants using *E. coli, Pseudomonas* sp., *Bacillus subtilis* and *B. cereus*. Gene transfer is very much dependent on the metabolic state of donor and recipient; in most cases optimal growth conditions have resulted in the highest rates of genetic exchange. Most studies have not altered the soil pH but acidic and alkaline soils have not been studied in detail.

Polluted soils

A limited number of studies have considered the selection of introduced genes within a noxious soil environment containing heavy metals, antibiotics or xenobiotics. Polluted soils could be detoxified or their physical and chemical properties improved by bacterial activity. Genetically engineered bacteria have been constructed to improve degradative and resistance characteristics for use in the field. Brokamp & Schmidt (19) reported plasmid transfer under selective conditions in long-term microcosms (180 days) with homogenized soil. The soil contained 2,2-dichloropropionate (DCPA) and a plasmid bearing the degradative gene (*dhlC*) was transferred from *Alcaligenes xylosoxidans* to the indigenous soil bacteria (identified as *Pseudomonas* and *Alcaligenes* species). The data indicated that horizontal transfer of degradative genes was more likely to occur under selective conditions than adaptation to DCPA under the conditions used. Recorbet *et al.* (135) added neomycin and kanamycin to soil in an attempt to provide selection for the *nptII* phosphotransferase resistance gene on the chromosome of *E. coli*. Homogenized sterile and non-sterile microcosms were used to study the survival of marked *E. coli*. The addition of kanamycin to the soil did not enhance the survival of the resistant *E. coli*, numbers of which declined rapidly in non-sterile soil. More than 10^4 indigenous soil bacteria were resistant to kanamycin but none tested contained the *nptII* gene.

DETECTION OF RECOMBINANTS

The detection of bacteria in natural environments and enumeration of rare genotypes requires sensitive and reliable methods. Isolation and culturing techniques have been used, combined with direct methods for detection *in situ*. There are two general approaches; firstly cell extraction followed by selective plating, and secondly direct enumeration *in situ*, for example immunofluorescence, scanning electron microscopy, *in situ* rRNA hybridization, and DNA/RNA extraction. Selective plating has been used for sensitive detection of transconjugants. It has not been possible to use direct methods as yet for detection of gene transfer in non-sterile soil. This is partly because of the problems in discriminating between recombinants and donors. Methods involving the analysis of DNA have allowed detection of new gene locations on the chromosome by restriction analysis of soil DNA (77) but the relative insensitivity of this method would make it difficult to detect recombinants. Holben *et al.* (77) seeded non-sterile soil with ~10^7 cells of an engineered strain of *Bradyrhizobium japonicum*, which could be differentiated from wild-type directly by restriction analysis of soil DNA. The polymerase chain reaction (PCR) could enhance the detection limit for specific sequences in soil and recombinants found by analysis of flanking sequences to the gene being tracked. However, to date PCR has not been used directly on soil DNA to detect gene transfer events.

Extraction

The simplest method for enumeration involves resuspension of the soil sample in an appropriate buffer/saline; the sample is then agitated to dislodge cells from the soil complex to enable isolation by serial dilution plating. Other physical methods for the concentration of bacteria extracted have been devised (7, 49); see also Chapter 5. Efficient extraction of filamentous bacteria such as streptomycetes is difficult due to the intricate association between soil particles and mycelia. Baecker & Ryan (6) recommended treatment of the soil by hammer-mill comminution and high-speed mixing in sterile distilled water containing Teepol (10 ppm). Ramsay (133) conducted experiments to compare the efficiency of several physical methods for extraction of bacteria from soil. Blending, shaking and ultrasonication were compared with direct counts in two different soils. No differences were observed in a sandy soil but ultrasonication recovered more cells from a silt loam soil. These findings indicate that extraction techniques need to be evaluated for each individual soil type if maximum efficiency of recovery is to be achieved.

As an alternative to physical dispersion of bacterial cells from soils methods have been developed which chemically disrupt the soil structure enabling isolation of bacterial fractions from the soil slurry. These methods used chemical desorbing agents such as Dowex 1A (102) and Chelex 100 (74); they also include relatively non-specific detergents (sodium cholate; sodium deoxycholate) to help break up cell aggregates and soil complexes containing cells. These methods can be used to concentrate the total microbial population or, by selective elutriation and differential centrifugation, be used to selectively isolate a specific size or type of organism (74). The spore extraction technique was used successfully by Herron & Wellington (74, 75) to detect lysogens in soil, and revealed the survival of donor and transconjugant spores in a fed-batch microcosm (35).

Selection of recombinants

The majority of gene transfer experiments carried out in soil have relied on selective isolation of inoculants, mainly using antibiotic resistance markers (see Ref. 34 for review). Antibiotics have proved useful for the selection of resistant inoculants although insensitive indigenous bacteria and fungi may still prove to be a problem at low dilutions. Selection of resistant phenotypes must prevent the possibility of *in vitro* gene transfer on plates selective for transconjugants (148).

Other methods for the detection of bacteria in environmental samples include light production with *luxAB* (113, 145), *xylE* (109) and *lacZY* (44). These genes may be used as markers to detect gene transfer if counter-selection of the donor is achieved by other methods. It may be possible to use a repressor gene which prevents transcription of the marker in the donor; if the gene is transferred into a new host then the marker may be expressed and detected. However, in non-sterile systems the indigenous populations of bacteria may interfere with detection and recovery of the recombinant population if antimicrobial agents are not used (159). Marker genes may also be used to detect the inocula without the need for biomass extraction and culture. Morgan *et al.* (109) detected cells released into lakewater by ELISA techniques, assaying the activity of 2,3-dioxygenase from cells of *E. coli* and *Pseudomonas putida* harbouring recombinant plasmids encoding a *xylE* marker gene (187). Detection limits using this system were around 10^3–10^4 cells per ml lakewater. The *xylE* system has also been applied for the detection of streptomycetes in soil (188) by using DNA extraction methods and PCR amplification of the *xylE* gene. The *lux* system has been applied to the direct detection of inocula in soil. Rattray *et al.* (134) estimated light output by luminometry from luminescent *E. coli* cells introduced into soil; detection limits using this method were around 10^3–10^4 cells per g soil.

Direct detection of DNA/rRNA sequences

Direct detection of introduced inocula in soil can be achieved by direct or indirect extraction of DNA/rRNA from environmental samples. Indirect extraction of nucleic acids involves the selective extraction of the bacterial community; the basis of these methods were originally described by Torsvik & Goksøyr (161) and Torsvik (160). Using repeated rounds of homogenization with phosphate buffer, SDS (77, 150) and differential centrifugation, as first described by Fægri *et al.* (49), it is possible to extract and concentrate the bacterial fraction of an environmental sample, usually soil or sediment. The cells can then be washed and lysed with lysozyme and SDS treatment and the DNA precipitated by standard protocols. Indirect lysis has been used to detect and quantify specific microbial populations in soil (77). Whole community DNA was probed by slot blot or Southern blot methods with *nptII* genes specific for *Bradyrhizobium japonicum*. Detection limits using this methodology were around 4.0×10^4 cells per g soil. Steffan & Atlas

(149) used indirect DNA extraction techniques combined with PCR amplification to enhance the detection of *Pseudomonas cepacia* AC1100 by dot blot analysis. The PCR amplification improved detection by a factor of approximately 10^3.

Direct extraction of DNA from soil was originally described by Ogram *et al.* (114) and involved treatment of the environmental sample with chemical and physical extractions. The object of the treatment was to lyse the bacterial cells *in situ* and subsequently extract the released DNA directly from the sample. Alkaline sodium phosphate buffer (114) is commonly used for the extraction of the DNA from the soil during cell lysis.

Physical and chemical methods which are used for cell lysis, based on the methods of Ogram *et al.* (114) are:

1. Detergent, e.g. sodium dodecyl sulphate (SDS).
2. Heat treatment (70°C).
3. Bead beating (Bead beater, Braun Melsungen).

Tsai & Olson (167) described a direct method of extraction of DNA from soil which differed from the Ogram method; in addition to SDS and lysozyme treatment a freeze–thaw regime was also used. Sayler *et al.* (142) used the direct lysis procedures to isolate DNA and RNA from soils. The DNA, after purification using hydroxyapatite columns, was susceptible to enzymic digestion and probed by slot blot analysis. Cresswell *et al.* (33) used two different direct DNA extraction techniques to isolate plasmid DNA from soil samples containing *Streptomyces violaceolatus* harbouring the multicopy plasmid, pIJ673 (179). It was demonstrated that spores of *S. violaceolatus* were resistant to SDS/heat lysis but susceptible to bead-beating lysis; the combination of the two methods enabled the cellular origin of the plasmid DNA to be determined. This method was applied to the study of a plasmid population in soil (35). A close correlation was found between levels of total plasmid DNA and numbers of donors and transconjugants. The latter were only detected where donor and recipient populations peaked at 17 days during a 60-day incubation (181).

Both direct and indirect DNA isolation procedures yield DNA in sufficient quantity to enable further purification for molecular analysis. However, there are several distinct differences in the quantity and quality of DNA recovered. Using indirect lysis, DNA is protected from contamination during the initial isolation of the bacterial fraction. This can prevent gross contamination of the DNA with humic acids (especially in soils with a high organic content). Once isolated the cells can be treated in various ways, e.g. washed, then resuspended in an appropriate buffer and so lysed in a more defined environment. DNA recovered after precipitation is cleaner than DNA from the direct lysis method and may require less purification for restriction (77). However, the amount of DNA recovered from soils and sediments can be significantly less than that recovered from direct lysis methods (150). This may reflect the efficiency of the extraction of the bacterial fraction from the sample.

Direct lysis protocols recover all types of available DNA from environmental samples, depending on the lysis conditions used. Both indirect and direct lysis procedures will contain DNA from non-viable and non-culturable bacteria. Direct lysis extracts will also contain extracellular and eukaryotic DNA as well as intracellular DNA.

Specific methods for the extraction of RNA have also been developed. rRNA sequence data obtained for target genera can be analysed to produce oligonucleotide probes which can bind to the appropriate DNA/RNA sequences. Such probes have been selected for a range of specificities. Hahn *et al.* (63) developed methods using indirect and direct isolation of rRNA from soil samples. Direct isolation of rRNA involved sonication of the soil sample in 7.5 M guanidine hydrochloride based on methodology originally used to recover RNA from root nodules (64); direct isolation yielded more RNA than that obtained by an indirect lysis procedure.

It is possible to use DNA/rRNA probing methods quantitatively. This depends on knowing the size of the DNA/rRNA probe and the copy number of the target sequence in the population. Radiolabelled probes used with control and sample DNA can be used for quantification with a Molecular Dynamics Phosphorimager.

Detection of inoculants in soil using scanning electron microscopy (SEM)

Few studies have applied SEM to the detection of soil bacteria. This is because in a non-sterile soil environment the detection of bacteria is very difficult. Clewlow *et al.* (30) and Wellington *et al.* (180) used SEM to investigate the growth cycle and differentiation of streptomycete inoculants in sterile unamended soil. These studies involved laborious sequential scanning of microcosm soil samples from different sample times. The results enabled the time scale to be determined for germination, mycelial development and genetic interactions *in situ* in sterile soil. This information was compared with results obtained using a spore-specific extraction method (74). The findings confirmed that germination occurred during the first 2 days of incubation and that sporulation began between 2 and 3 days. These observations were also used to formulate a mathematical model of plasmid transfer between streptomycete inoculants in soil (30). The first part of the model assumed plasmid transfer occurred only within the first two days. After this time the increase in transconjugant numbers was attributed to sporulation only. The model predictions fitted the experimental data.

SURVIVAL OF ENGINEERED BACTERIA IN THE SOIL AND TRANSFER OF NEW GENES

New genes may be introduced into indigenous soil populations of bacteria but microcosm studies have indicated that this will occur at a very low frequency in the field. One of the major factors to be determined is the survival of introduced bacteria which might act as donors. The second consideration is the likelihood that the DNA would not survive in the indigenous population if it has no selective advantage. Chemostat studies have shown that the possession of non-chromosomal genetic elements (e.g. plasmids, transposons and temperate bacteriophage) reduced (69, 82, 84), enhanced (69) or had no effect on the fitness of a host bacterium (82). It thus appears that the effect of extra DNA in such systems is dependent on the nature and products of the DNA rather than the DNA burden itself. Some species of microorganisms (genetically engineered or non-genetically engineered) persist following their introduction into the environment. Other species introduced into the same environment in large numbers die out rapidly. Many observations have been made on the decline or persistence of different bacterial species (Tables 6.1, 6.3, 6.5 and 6.7), but it has not always been possible to determine which factors are most important in determining survival in natural ecosystems (2). The competitiveness of non-genetically engineered inoculants in the environment has been best studied in *Rhizobium* with respect to environmental conditions (Table 6.7), although many other microorganisms have also been studied. Some conditions exert a negative effect, others a positive and some have no effect at all on the survival. The introduction of DNA as plasmids, transposons or chromosomal rearrangements, or the deletion of DNA, affects the survival but no consistent pattern can be found (Tables 6.2, 6.4, 6.6, 6.7 and 6.8). The following review of survival studies involving selected bacterial inoculants demonstrates that members of genera indigenous to soil survive better than exotic groups such as enteric inoculants. Many of the studies in Tables 6.1–6.8 are not directly comparable because of the differences in types of microcosm and field plots used. Factors such as time of inoculation may be critically important in field experiments whereas in microcosms this is irrelevant.

Survival and gene transfer between *Bacillus* species

Many *Bacillus* species are indigenous to soil and their ability to sporulate should enhance survival. Table 6.1 illustrates the importance of the

Table 6.1 Survival of *Bacillus* in the environment: effect of environmental conditions.

Organism	Environment*	Condition*	Effect on survival†	Reference
Bacillus	Soil	Nutrient availability, moisture	+ +	183
Bacillus (asporogenous)	Lakewater	Amendment	+	70

* Experimental date from microcosms or field sites.
† + = enhanced survival.

environmental nutrient status for the growth and thus improved survival of *Bacillus* inoculants in soil and water. Few studies have investigated the effect of mutation or introduction of cloned DNA on the survival of inoculants. Van Elsas *et al.* (170) reported that the introduction of plasmid pFT30 into *Bacillus subtilis* did not affect the survival of this strain in soil. The only documented environmental release involving an engineered *Bacillus* species used marked strains of *Bacillus thuringiensis* which were released by Ecogen in 11 US states in field plots varying between 0.1 and 7.5 acres per site.

Table 6.2 Survival of enteric and other bacteria in the environment: effect of environmental conditions.

Organism	Environment	Condition	Effect on survival*	Reference
Escherichia Salmonella	Faeces buried in soil	Time	− −	158
Escherichia	River-water	Competition	−	51
Escherichia	Water	Seawater Freshwater +Humic acids	− − +	38
Escherichia	Soil	Nutrient availability, Competition	+ +	84
Faecal indicator bacteria	Estuary	Time	−	119
Klebsiella Gram-negatives	Lakewater	Time	−	147
Klebsiella Escherichia	Tropical marine	Time	0/−	96
Pseudomonas Escherichia Klebsiella Micrococcus Arthrobacter Rhizobium Bacillus (asporogenous)	Lakewater	Amendment	+ + + 0 0 + −	70
Pseudomonas Salmonella Escherichia Klebsiella	Sediment	Time	− − − −	22

* + = enhanced survival; − = reduced survival; 0 = no effect on survival.

Intrageneric gene transfer has been observed between bacilli in sterile amended soil and non-sterile soil amended with bentonite (171), but was undetectable in wheat rhizosphere (172). *Bacillus* is naturally transformable (99, 100). The only report of the transformation of auxotrophic and resistance markers between *Bacillus subtilis* strains was in sterile amended soil (61, 62). As both donor and recipient were free from plasmids and transducing phage this transfer was attributed to transformation. The addition of DNase did not prevent the process, as would be expected if transformation was involved. However, the inactivity of the DNase was assumed to be due to either inactivation of the enzyme or DNA protection by clay particles. There is no information as to whether *Bacillus* species can be transduced in natural soil.

Survival and gene transfer between enteric, plant pathogenic and other related bacteria

E. coli is not a natural inhabitant of soil. However, experiments investigating its gene transfer capability have relevance to the safety and potential escape of genes from laboratory strains. A number of studies have monitored survival in soil and particular attention has been paid to aquatic environments where the contamination of potable waters is assessed by monitoring faecal coliforms (Table 6.2). Again as with other introduced strains the nutrient status of the environment is important in determining the extent of survival. In soil, enhanced nutrient status may improve survival but in the presence of indigenous microflora enteric bacteria will be outcompeted (Table 6.2).

A wide range of experiments have been undertaken to monitor gene transfer between enteric bacteria and also to determine if genes can be transferred to the indigenous microflora. Table 6.3 illustrates that enteric bacteria, mainly *E. coli* strains containing recombinant plasmids, were capable of surviving in a range of environments. The introduced DNA in most cases appeared to have no effect on survival.

E. coli can survive for at least 3 weeks in soil (88) and mobilize a plasmid to indigenous bacteria (73). Conjugation takes place in soil within many genera. The first report was by Weinberg & Stotzky (177), who demonstrated conjugation between strains of *E. coli* in sterile soil (both plasmids and chromosomal DNA were transferred). It was found that the addition of montmorillonite clay to the system enhanced the frequency of conjugation, presumably through enhanced adhesive properties of the soil and stabilization of the mating pair. Since then *E. coli* conjugation has been demonstrated in sterile amended soil (162, 163) and in non-sterile soil (88, 164), where the process was shown to be pH dependent. Less information on intergeneric conjugation in soil is available. It has been demonstrated in sterile soil from *E. coli* to *Rhizobium*, where transfer frequency was shown to be dependent on a number of abiotic factors (136). Conjugation was also demonstrated in sterile soil (71) between *E. coli* and *Enterobacter* and between *Enterobacter*

Table 6.3 Survival of enteric and plant pathogenic bacteria in the environment: effect of cloned or mutated DNA.

Organism	Environment*	Cloned or mutated DNA†	Effect on survival‡	Reference
Agrobacterium *Xanthomonas* *Erwinia*	Phytopathogenicity	pRD1 (plasmid)	+ +/− +/−	87
Enterobacter	Rhizosphere	pRD1 (plasmid)	S	83
Erwinia	Pondwater	km^r	0	143, 144
Erwinia	Soil	km^r	0/+	116
Escherichia	Soil	Plasmids	+	41
Escherichia	Soil	pBR322 (plasmid containing *Drosophila* gene)	0	42
Escherichia	Air	ColEl::Tn5 (plasmid)	0/+	103
Escherichia	Intestine	ab^r (mut^n)	−/0/+	115
Escherichia	Intestine	pBR322 (plasmid)	+	92
Escherichia *Pseudomonas*	Lakewater	Plasmid	− 0	5
Escherichia	Riverwater Soil	Plasmids	0 +	27
Escherichia	Seawater	pBR322 pUC8 (plasmids)	0 0	23
Escherichia	Well-water	R388 pRO101 (plasmids)	− −	24
Pseudomonas		R388 pRO101	− 0	
Alcaligenes	Lakewater	pBR60 (chlorobenzoate degrading plasmid)	+/−	55
Azospirillum	Soil	Tn5	S	12

* Experimental data from microcosms or field sites.
† rif^r = rifampicin resistance; km^r = kanamycin resistance; ab^r = antibiotic resistances; mut^n = chromosomal mutation; TD, doubling time.
‡ + = enhanced survival; − = reduced survival; 0 = no effect on survival; S = organism survived.

and *Pseudomonas*. Meanwhile, Top *et al.* (159) were able to show conjugation in sterile and non-sterile amended soil between *E. coli* and *Alcaligenes*. Transfer of plasmid pFL67-2 (based on the promiscuous plasmid RSF1010) from *E. coli* to indigenous *Pseudomonas fluorescens* strains was shown in natural soil by Henschke & Schmidt (73). Gene transfer can also take place *in planta*, e.g. from *E. coli* to *Pseudomonas* (91); between *Klebsiella* strains on radish plants (157) and has been reviewed by Farrand (50). The first report of transduction in the environment was by Baross *et al.* (11), who showed the transduction of agarase genes between strains of *Vibrio parahaemolyticus* in the guts of oysters in sterile seawater. The possible mechanisms of this process in the environment have been reviewed by Kokjohn (86). In addition to *Pseudomonas* (see below), many other genera undergo natural transformation, e.g. epilithic *Acinetobacter* (137) and *Vibrio* (53, 81). As with transformation, transduction in soil is also a poorly studied process. Germida & Khachatourians (57) raised a transducing lysate from a Tn*10*-containing strain of *E. coli* and used this to transduce an auxotrophic and tetracycline-sensitive strain *in situ*. *E. coli* was also used to demonstrate phage-mediated gene transfer in soil by Zeph *et al.* (191), although this was not transduction but a phage conversion (10). A derivative of phage P1 (containing Tn*501*) was used to create lysogens of a marked *E. coli* strain; this was then inoculated into soil with a marked non-lysogenic recipient. The appearance of the resistance markers of Tn*501* in the recipient showed the phage had been induced from the donor and reinfected the recipient. The transfer was confirmed by the use of nucleic acid probing (192).

Survival and gene transfer among *Pseudomonas* species

Pseudomonads are indigenous in soil and water, with many species able to colonize plants; members of this genus often show resistance to adverse environmental conditions such as heavy metals and agricultural antibiotics. Using metal resistance as a marker, Pickup (120) found that 60% of copper tolerant strains isolated from lakewater carried at least one plasmid. The survival of introduced strains depends on the time of introduction and the extent of competition from indigenous pseudomonads; many species are specialized colonizers of plant surfaces and would not survive in large numbers in the bulk soil (Table 6.4). The introduction of recombinant DNA on plasmids or the chromosome did not appear to adversely effect survival although the studies summarized in Table 6.5 indicated that certain plasmids and the transposon Tn*5* may reduce fitness in the soil.

A number of field trials have been conducted with engineered strains, including *Pseudomonas aureofaciens* (*lacZY*, Monsanto), *P. fluorescens* (*lacZY*), *P. putida* (siderophore deletion) and *P. syringae* (Ice⁻, University of California).

Much attention has focused on the safe use of recombinant pseudomonad inoculants and this is illustrated in Table 6.5, where studies of the fate of

Table 6.4 Survival of *Pseudomonas* in the environment: effect of environmental conditions.

Organism	Environment	Condition	Effect on survival*	Reference
Pseudomonas/ actinomycetes	Soil	Location	S	108
Pseudomonas (Ice nucleation ability)	Soil Water Snow	Time	– – –	59
Pseudomonas/ Arthrobacter	Soil	Nutrient availability Moisture Competition	+/0 –/0 0/0	90

* + = enhanced survival; – = reduced survival; 0 = no effect on survival; S = strain-specific survival.

engineered strains in soil and water are summarized. The fate of recombinant plasmids has received most attention. Genther *et al.* (56) reported that transfer frequencies between a *Pseudomonas* donor and natural isolate recipients were higher on plates than in liquid culture. An unusual observation was made by McClure *et al.* (105), who, after inoculating a non-conjugative plasmid-bearing strain of *Pseudomonas* into an activated sludge unit, subsequently re-isolated the inoculum to find that it had acquired indigenous plasmids that had the ability to mobilize the original plasmid to recipient strains *in vitro*. Conjugation also occurs extensively in the epilithon (the slime layer covering stones in aquatic habitats). Naturally isolated, mercury resistance plasmids that were able to transfer over a wide range of environmentally significant conditions (138) were used to demonstrate gene transfer between *Pseudomonas* strains in the epilithon (8, 9, 40, 54, 139). Conjugation in the terrestrial environment has been detected between pseudomonads in sterile soil (165) and in a sterile pea-vermiculite mixture, rhizosphere and phyllosphere (85). It has also been demonstrated in non-sterile bulk soil and wheat rhizosphere, but was enhanced in the rhizosphere or when bulk soil was amended with bentonite or nutrients (172, 173). Extracellular DNA is produced and excreted by aquatic bacteria (117, 118); thus, along with DNA released from lysed bacteria, there is a potential for transformation of aquatic microorganisms in the aquatic ecosystem (reviewed in Ref. 151). If this DNA can adsorb to sand or sediment particles it then becomes much more resistant to the action of exonucleases (1, 97, 98). Natural transformation of pseudomonads occurs *in vitro* (25, 153) and also when attached to solids *in situ* (101, 154), although Stewart *et al.* (155) found that exogenous DNA is only biologically active when the solids (autoclaved

Table 6.5 Survival of *Pseudomonas* in the environment: effect of cloned or mutated DNA.

Organism	Environment*	Cloned or mutated DNA†	Effect on survival‡	Reference
Escherichia *Pseudomonas*	Lakewater	Plasmid	− 0	5
Pseudomonas	Soil	Plasmid (pollutant degrading)	S	47, 48
Pseudomonas	Activated sludge	pD10 (plasmid)	0	105
Pseudomonas/ *Klebsiella*	Drainage water	Plasmids	−	175
Pseudomonas	Groundwater	pWWO RK2 (plasmids)	0 0	80
Pseudomonas	Phylloplane	Ice minus	+	95
Pseudomonas	Soil	pBS3 (Kelthane degrading plasmid)	+	58
Pseudomonas	Soil	pLAFR3 pJE8	0 0	189
Pseudomonas	Soil	pRO101, pRO103 (herbicide degrading plasmids)	0	146
Pseudomonas	Soil	Plasmid (herbicide-degrading plasmid)	0	132
Pseudomonas	Soil	rif^r (mut^n)	0/−	31
Pseudomonas *Bacillus*	Soil	Tn*5* pFT30 (plasmid)	0 0	170
Pseudomonas	Soil	Tn*5*/Tn*5*::*tox*	− −	174
Pseudomonas	Tropical marine	megaplasmid	0	37

* Experimental data from microcosms or field sites.
† rif^r =rifampicin resistance; mut^n = chromosomal mutation.
‡ +=enhanced survival; − = reduced survival; 0 = no effect on survival.
S = strain-specific survival

sediment) were first saturated with non-homologous DNA. This process would seem to be ecologically important to the *Pseudomonas* donor strain involved as Stewart *et al.* (152) showed the process to be active and plasmid independent. Transduction and co-transduction have been shown between strains of *Pseudomonas* in lakewater (111, 140, 141). Morrison *et al.* (111) found that, as well as transduction of a recipient by an inoculated free phage lysate, lysogenic donors could be induced *in situ* and subsequently that transducing phage particles, carrying a chromosomal marker gene, could reinfect a non-lysogenic recipient. Later work by Saye *et al.* (140) only detected transduction of plasmids in a system which used a lysogenic recipient strain and a non-lysogenic plasmid donor. Saye *et al.* (141) found no significant difference in the transduction of chromosomal DNA and plasmid DNA *in vitro*, although higher frequencies of plasmid transduction were found *in situ*. This study also showed that chromosomal markers could be co-transduced *in situ* and that transductants of lysogenized strains were obtained 10–100-fold more frequently than from non-lysogenic parents.

Survival and gene transfer within *Rhizobium* species

Rhizobium species reside in the soil prior to interacting with plant roots and subsequently producing a nutrient-enriched environment for growth. They survive in soil but are unlikely to be active. Their ability to introduce fixed nitrogen to the soil ecosystem has made them candidates for long-term inoculation studies. As Table 6.6 shows introduced strains can survive in soil although predation may reduce numbers, but the nodule provides conditions suitable for growth. Rhizobia showed reduced survival in soils with increased water content (Table 6.6).

Several field releases of engineered rhizobia have taken place using marked nitrogen-fixing strains of *Bradyrhizobium japonicum* and *Rhizobium melliloti*. This has increased interest in the extent of genetic interactions among rhizobia in soil and the fate of introduced DNA has been studied for a range of species in differing soil types (Table 6.7). Members of the genus *Rhizobium* are able to conjugate in sterile alfalfa nodules (127), and are able to receive plasmid DNA from *E. coli*, where transfer frequency was shown to be dependent on a number of abiotic factors (136). Neither transformation nor transduction have been demonstrated in the environment within this genus.

Survival and gene transfer among *Streptomyces* in soil

Streptomycetes are indigenous in soil and play an important role in the decomposition of plant and animal remains. They produce spores which improve resistance to desiccation although asporogenous strains can be isolated from soil and mutants have survived in non-sterile soil (180).

Streptomyces species are important commercially for the production of

Table 6.6 Survival of *Rhizobium* in the environment: effect of environmental conditions.

Enrivonment	Condition	Effect on survival	Reference
Bean rhizosphere	Protozoan inhibition	+	131
Nodule	Acidity	S	45, 46
Soil	Competition	S	107
Nodule	Competition	S	79
Soil	Predation	−	125
Soil	Clay content	+	67
	Raised water content	−	
Soil	Protection from predation by clay	+	66
Liquid culture	Protection from predation by clay	+	68
Soil	Raised water content	−	124
Soil	Raised water content	−	123
Soil	Temperature	+	18
	Moisture	−	
Soil	Time	−	36
Soil/nodule	Inoculum size	+	104
Soil/peat	Soil type	S	106

* + = enhanced survival; − = reduced survival; S = strain-specific survival.

Table 6.7 Survival of *Rhizobium* in the environment: effect of cloned or mutated DNA.

Enrivonment	Cloned or mutated DNA*	Effect on survival†	Reference
Desert soil	Tn5	0	121
Soil	Tn5	−	76
Sand/soil	Tn5	0	26
Soil	ab^r (mut^n)	0	89
Root nodule	ab^r (mut^n)		168
	(Nod^n)	−	
	(N$_2$ fix^n)	0/−	
	(TD)	+	
Root nodule (nodulation)	rif^r (mut^n)	−	20
	pTA2 (plasmid)	+	
Root nodule (nodulation)	rif^r (mut^n)	−	93

* rif^r = rifampicin resistance; ab^r = antibiotic resistances; mut^n = chromosomal mutation; TD = doubling time; Nod^n = nodulation.
† + = enhanced survival; − = reduced survival; 0 = no effect on survival.

antibiotics and enzymes. Large amounts of live mycelium are disposed of in the environment as waste from antibiotic broths. No deliberate releases of engineered strains have occurred but disposal of recombinant industrial species will need careful monitoring. Such strains may survive in soil (Table 6.8) but mutated laboratory strains did not survive well in competition with the indigenous microflora (Table 6.8; Ref. 35). Wang et al. (176) reported the survival of a *Streptomyces lividans* strain containing recombinant plasmids in both sterile and non-sterile soil. The plasmids did not appear to reduce survival and inoculants were still detected after 10 months.

Table 6.8 Survival of *Streptomyces* in the environment: effect of cloned or mutated DNA.

Enrivonment	Cloned or mutated DNA*	Effect on survival†	Reference
Soil	Plasmids	0	176
	Mut^n	0	
Soil	plasmid-mut^n	–	179
			35

* mut^n = chromosomal mutation.
† 0 = no effect on survival; – = reduced survival.

   Streptomycetes are capable of exchanging genetic information *in situ* by means of conjugation. Transformation and transduction have not yet been demonstrated. It is thought, however, that gene transfer by these mechanisms would be rare, due to the lack of transducing phage (29), the high level of restriction in this genus (32) and that *Streptomyces* species are not naturally competent for transformation. The first reports of conjugation in sterile soil were made by Rafii & Crawford (128) and Wellington et al. (178), who showed that derivatives of the self-transmissible, multi-copy plasmid, pIJ101, could be transferred and mobilized between strains of S. *lividans* and other streptomycetes. Bleakley & Crawford (15) reported that the number of transconjugants in sterile soil was increased by amendment of the soil and was optimal in soil with a lower water content. Rafii & Crawford (130) detected conjugation in liquid culture and suggested that *Streptomyces* may be capable of mating in aqueous systems. Using a system based on S. *lividans* and a pIJ101 derivative it was found that these organisms could conjugate intra- and interspecifically in non-sterile soil (35, 179, 180). This system was used for the development of a mathematical model based on the Pearl–Verhulst logistic equation, which was capable of accurately predicting plasmid transfer, population dynamics and response to nutrient and moisture limitation (30). The latter was important in batch microcosms where both nutrients and moisture became limiting to growth and phage activity as illustrated in Fig.

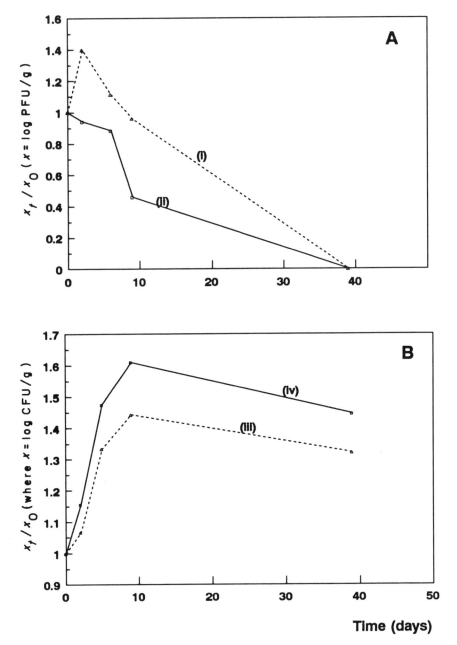

Fig. 6.1 Host and phage interactions in a closed soil microcosm containing sterile, amended soil. A, survival of KC301 phage (i) with host and (ii) without host. B, survival of *Streptomyces lividans* TK24 (iii) with phage and (iv) without phage. Phage numbers expressed as $\log_{10}$ PFU (plaque forming units) per gram. X_t = log count at time t; X_o = log count at time 0.

6.1B. A rapid increase in total counts (spores plus mycelium) occurred within the first 10 days; after this period no further spore germination was observed and counts remained static. Actinophage have been used to study the fate of phage-borne genes (74) and in a batch microcosm, as illustrated in Fig. 6.1A, the activity of the inoculant, *S. lividans*, correlated with a burst in numbers of the recombinant phage, KC301. No further increases in phage numbers indicated that the host was not active or susceptible to infection; only active mycelium is susceptible. It is possible to conclude from Figs 6.1A and 6.1B that germinating spores and young mycelium were only present for a very short period of time, probably just the first 2 days, after which time phage numbers declined rapidly. Phage did not replicate in the absence of host. The presence of KC301 did not significantly affect host numbers of colony forming units recovered. The latter is not surprising because the efficiency of infection in this system was calculated to be approximately 10^{-6}–10^{-7} with the phage inoculum density indicated in Fig. 6.3.

Closed microcosms, therefore, have limited value for more detailed population studies as only one round of spore germination occurs. An open microcosm was devised which resembled a fed-batch system where 50% of the spent soil was removed and substituted with fresh amended soil every 15 days (Figs 6.2 and 6.3). Repeated rounds of spore germination (Figs 6.2A

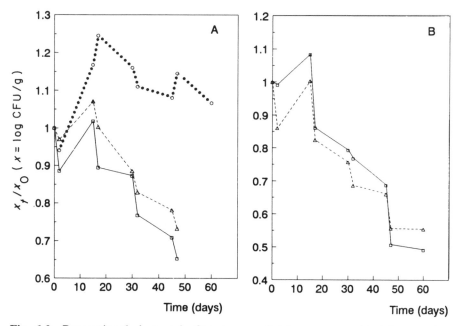

Fig. 6.2 Proportional changes in *Streptomyces lividans* spore and total counts in a fed-batch soil microcosm. A, total counts; B, spore counts. *S. lividans* TK24 (□); *S. lividans* TK24, plus KC301 (△); indigenous streptomycetes (O). $X_t = \log_{10}$ count at time t; $X_o = \log_{10}$ count at time 0.

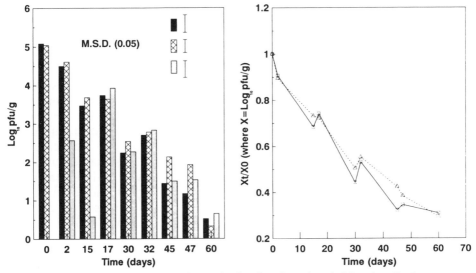

Fig. 6.3 Population dynamics of a marked actinophage in a fed-batch soil microcosm. ▮, KC301 phage marked with *tsr* gene (△); ▨, KC301 plus TK24 host *Streptomyces lividans* (□); ▯, indigenous actinophage.

and 6.2B) and phage infections (Figs 6.3A and 6.3B) were evident. The marked phage, KC301, showed replication after fresh amendments of the soil, correlating with spore germination events occurring around day 2, 17, 32 and 47. The decline in the inoculant populations has masked the troughs expected in spore counts other than that seen at day 2 (Fig. 6.2B). Indigenous actinophage, which were counted using *S. lividans* as host lawns, were not detected after drying and wetting of non-sterile soil (day 0, Fig. 6.3) but soon increased in numbers as both the indigenous streptomycetes and inoculant strains germinated (Fig. 6.2A). KC301 showed a significant decline within this population but so too did indigenous actinophage (those plating on *S. lividans*), perhaps in response to the demise of the *S. lividans* population.

The poor survival of laboratory strains is shown in Figs 6.2A and 6.2B. Indigenous streptomycetes did not decline but fluctuated around 10^7 (total count). Gene transfer was found to be dependent on densities of donors and recipients as transconjugants were only detected at significant levels at day 15 and 17; both populations peaked at day 15. No plasmid transfer to the indigenous population was detected under these conditions although even at low dilutions rare indigenous transconjugants would probably have been outnumbered by the inoculated donors. To detect such events counter-selection of the donor is necessary and selection of transferred genes is useful to allow such transconjugants to increase in numbers.

Open systems, as illustrated here, allow investigation of population dynamics over extended periods of time and can increase the chances of detecting transfer events and enable studies of the effects of selection on recombinants (35).

CONCLUSIONS

Data from microcosm experiments have shown that introduced bacteria can survive and transfer their genes both to seeded recipients and to indigenous populations. Behaviour in the field is difficult to predict; so far the experimental approach has failed to provide evidence for gene transfer. It is likely that such events are rare and poor survival and growth in natural soil may reduce numbers of recombinants to below detection limits. Too few field experiments have been done to compare these data with the frequencies gathered from model systems, where conditions for growth are near the optimal habitat likely to occur in soil. The use of model systems has enabled all modes of gene transfer to be demonstrated in a soil environment and factors limiting frequencies were usually related to nutrient availability. One of the most interesting aspects which has received less attention is the role of selection on the survival and spread of introduced genes in indigenous populations of bacteria. Undoubtedly the selective value of a gene may become important for its spread once selection is exerting an effect on a natural population. The considerable selective pressure for antibiotic resistance in clinical environments is a clear example of this. The extent to which such selection pressure could occur in soil is unknown but the extreme heterogeneity of soil may reduce selective effects. Factors such as pollution, which serve to distort the environment, may have an impact on gene transfer by reducing habitat variation.

ACKNOWLEDGEMENTS

We are grateful to the Natural Environment Research Council (Grant GST/ 02/191(B)), Science and Engineering Research Council (studentship PRH) and Commission of the European Communities (BAP 0378-UK, C11 0545-UK) for sponsorship of some of the work reported in this chapter.

REFERENCES

1. Aardema, B.W., Lorenz, M.G. & Krumbein, W.E. (1983) *Applied and Environmental Microbiology* **46**, 417–420.
2. Alexander, M. (1981) *Annual Reviews of Microbiology* **35**, 113–133.
3. Anderson, J.M. (1987) In *Ecology of Microbial Communities*, Eds Fletcher, M., Gray, T.R.G. & Jones, J.G. Society for General Microbiology Symposium No. 41. Cambridge University Press, Cambridge, pp. 125–145.
4. Anderson, J.M. & Ineson, P. (1982) *Soil Biology and Biochemistry* **14**, 415–416.
5. Awong, J., Bitton, G. & Rasul, G. (1990) *Applied and Environmental Microbiology* **56**, 977–983.
6. Baecker, A.A.W. & Ryan, K.C. (1987) *South African Journal of Plant and Soil* **4**, 165–170.

7. Bakken, L.R. (1985) *Applied and Environmental Microbiology* **49**, 1482–1487.
8. Bale, M.J., Fry, J.C. & Day, M.J. (1987) *Journal of General Microbiology* **133**, 3099–3107.
9. Bale, M.J., Fry, J.C. & Day, M.J. (1988) *Applied and Environmental Microbiology* **54**, 972–978.
10. Barksdale, L. & Arden, S.B. (1974) *Annual Reviews of Microbiology* **28**, 265–299.
11. Baross, J.A., Liston, J. & Morita R.Y. (1974) In *International Symposium on Vibrio parahaemolyticus*, Eds Fujimo, T., Sakagnehr, G., Sakazaki, R. & Takeda, Y. Saikon Press, Tokyo, pp. 129–137.
12. Bentjen, S.A., Fredrickson, J.K., Van Voris, P. & Li, S.W. (1989) *Applied and Environmental Microbiology* **55**, 198–202.
13. Bishop, D.H.L., Entwistle, P.F., Cameron, I.R., Allen, C.J. & Posse, R.D. (1988) In *The Release of Genetically-Engineered Micro-organisms*, Eds Sussman, M., Collins, C.H., Skinner, F.A. & Stewart-Tull, D.E. Academic Press, London, pp. 143–206.
14. Blair, J.M., Crossley, Jr, D.A. & Rider, S. (1989) *Soil Biology and Biochemistry* **21**, 507–510.
15. Bleakley, B.H. & Crawford, D.L. (1989) *Canadian Journal of Microbiology* **35**, 544–549.
16. Bolton, H. Jr., Fredrickson, J.K., Bentjen, S.A., Warkman, D.J., Shumei, W., Li, S.W. *et al.* (1991) *Microbial Ecology* **21**, 163–173.
17. Bolton, H. Jr., Fredrickson, J.K., Thomas, J.M., Warkman, D.J., Bentjen S.A. & Stuart, J.L. (1991) *Microbial Ecology* **21**, 175–189.
18. Boonkerd, N. & Weaver, R.W. (1982) *Applied and Environmental Microbiology* **43**, 585–589.
19. Brokamp, A. & Schmidt, F.R.J. (1991) *Current Microbiology* **22**, 299–306.
20. Bromfield, E.S.P., Lewis, D.M. & Barran, L.R. (1985) *Journal of Bacteriology* **164**, 410–413.
21. Burns, R.G. (1988) In *Handbook of Laboratory Model Systems for Microbial Ecosystems Vol II*, Ed. Wimpenny, J.W.T. CRC Press, Boca Raton, Florida, pp. 51–98.
22. Burton, G.A., Gunnison, D. & Lanza, G.R. (1987) *Applied and Environmental Microbiology* **53**, 633–638.
23. Byrd, J.J. & Colwell, R.R. (1990) *Applied and Environmental Microbiology* **56**, 2104–2107.
24. Caldwell, B.A., Ye, C., Griffiths, R.P. & Moyer, C.L. (1989) *Applied and Environmental Microbiology* **55**, 1860–1864.
25. Carlson, C.A., Pierson, L.S., Rosen, J.J. & Ingraham, J.L. (1983) *Journal of Bacteriology* **153**, 93–99.
26. Catlow, H.Y., Glenn, A.R. & Dilworth, M.J. (1990) *Soil Biology and Biochemistry* **22**, 331–336.
27. Chao, W.L. & Feng, R.L. (1990) *Journal of Applied Bacteriology* **68**, 319–325.
28. Chapman, S.J. & Gray, T.R.G. (1986) *Soil Biology and Biochemistry* **18**, 1–4.
29. Chater, K.F. (1986) In *The Bacteria, Vol IX: The Antibiotic Producing Streptomyces*, Eds Queener, S.W. & Day, L.E. Academic Press, Orlando, pp. 119–158.
30. Clewlow, L.J., Cresswell, N. & Wellington, E.M.H. (1990) *Applied and Environmental Microbiology* **56**, 3139–3145.
31. Compeau, G., Al-Achi, B.J., Platsouka, E. & Levy, S.B. (1988) *Applied and Environmental Microbiology* **54**, 2432–2438.

32. Cox, K.C. & Baltz, R.H. (1984) *Journal of Bacteriology* **159**, 499–504.
33. Cresswell, N., Saunders, V.A. & Wellington, E.M.H. (1991) *Letters in Applied Microbiology* **13**, 193–197.
34. Cresswell, N. & Wellington, E.M.H. (1992) In *Genetic Interactions Among Microorganisms in the Natural Environment*, Eds Wellington E.M. & van Elsas, J.D. Pergamon Press, Oxford, pp. 59–82.
35. Cresswell, N., Herron, P.R., Saunders, V.A. & Wellington, E.M.H. (1992) *Journal of General Microbiology* **138**, 659–666.
36. Crozat, Y., Cleyet-Marel, J.C., Giraud, J.J. & Okatan, M. (1982) *Soil Biology and Biochemistry* **14**, 1401–1405.
37. Cruz-Cruz, N.E., Toranzos, G.A., Ahearn, D.G. & Hazen, T.C. (1988) *Applied and Environmental Microbiology* **54**, 2574–2577.
38. Davies, C.M. & Evison, L.M. (1991) *Journal of Applied Bacteriology* **70**, 265–274.
39. Dawes, E.A. (1989) In *Bacteria in Nature: Vol. 3: Structure, Physiology and Genetic Adaptability*, Eds Poindexter, J.S. & Leadbetter, E.R. Plenum Press, New York, pp. 67–187.
40. Day, M.J., Bale, M.J. & Fry, J.C. (1988) In *Safety Assurance for Environmental Introductions: NATO ASI Series, Vol. G18*, Eds Fiksel, J. & Covello, V.T. Springer-Verlag, Berlin, pp. 181–197.
41. Devanas, M.A. & Stotzky, G. (1986) *Current Microbiology* **13**, 279–283.
42. Devanas, M.A., Rafaeli-Eshkol, D. & Stotzky G. (1986) *Current Microbiology* **13**, 269–277.
43. Dijkstra, A.F., Govaert, J.M., Scholten, G.H.N. & van Elsas, J.D. (1987) *Soil Biology and Biochemistry* **3**, 351–352.
44. Drahos, D.J., Hemming, B.C. & McPherson, S. (1986) *Biotechnology* **4**, 439–444.
45. Dughri, M.H. & Bottomley, P.J. (1983) *Applied and Environmental Microbiology* **46**, 1207–1213.
46. Dughri, M.H. & Bottomley, P.J. (1984) *Soil Biology and Biochemistry* **16**, 405–411.
47. Dwyer, D.F., Rojo, F. & Timmis, K.N. (1988) In *Release of Genetically Engineered Microorganisms*, Eds Sussman, M., Collins, C.H., Skinner, F.A. & Stewart-Tull, D.E. Academic Press, London, pp. 77–88.
48. Dwyer, D.F., Rojo, F. & Timmis, K.N. (1988) In *Risk Assessment For Deliberate Release*, Ed. Klingmüller, W. Springer-Verlag, Berlin-Heidelberg, pp. 100-109.
49. Fægri, A., Torsvik, V.L. & Goskøyr, J. (1977) *Soil Biology and Biochemistry* **9**, 105–112.
50. Farrand, S.K. (1989) In *Gene Transfer in the Environment*, Eds Levy, S.B. & Miller, R.V. McGraw Hill, New York, pp. 261–285.
51. Flint, K.P. (1987) *Journal of Applied Bacteriology* **63**, 261–270.
52. Fredrickson, J.K., Bentjen, S.A., Bolton H. Jr., Li, S.W. & van Voris, P. (1989) *Canadian Journal of Microbiology* **35**, 867–873.
53. Frischer, M.E., Thurmond, J.M. & Paul, J.H. (1990) *Applied and Environmental Microbiology* **56**, 3439–3444.
54. Fry, J.C. & Day, M.J. (1990) In *Bacterial Genetics in Natural Environments*, Eds Fry, J.C. & Day. M.J. Chapman & Hall, London, pp. 81–88.
55. Fulthorpe, R.A. & Wyndham, R.C. (1989) *Applied and Environmental Microbiology* **55**, 1190–1194.

56. Genther, A.J., Chatterjee, P., Barkay, T. & Bourquin, A.W. (1988) *Applied and Environmental Microbiology* **54**, 115–117.
57. Germida, J.J. & Khachatourians, G.G. (1988) *Canadian Journal of Microbiology* **34**, 190–193.
58. Golovleva, L.A., Pertsova, R.N., Boronin, A.M., Travkin, V.M. & Kozlovsky, S.A. (1988) *Applied and Environmental Microbiology* **54**, 1587–1590.
59. Goodnow, R.A., Harrison, M.D., Morris, J.D., Sweeting, K.B. & Laduca, R.J. (1990) *Applied and Environmental Microbiology* **56**, 2223–2227.
60. Gould, W.D., Bryant, R.J., Trofymow, J.A., Anderson, R.V., Elliott, E.T. & Coleman, D.C. (1981) *Soil Biology and Biochemistry* **13**, 487–492.
61. Graham, J.B. & Istock, C.A. (1978) *Molecular and General Genetics* **166**, 287–290.
62. Graham, J.B. & Istock, C.A. (1979) *Science* **204**, 637–639.
63. Hahn, D., Kester, R., Starrenburg, M.J.C. & Akkermans, A.D.L. (1990) *Archives of Microbiology* **154**, 329–335.
64. Hahn, D., Starrenburg, M.J.C. & Akkermans, A.D.L. (1990) *Applied and Environmental Microbiology* **56**, 1342–1346.
65. Hartel, P.G., Williamson, J.W. & Schell, M.A. (1990) *Soil Science Society of America Journal* **54**, 1021–1025.
66. Heijnen, C.E., Van Elsas, J.D., Kuikman, P.J. & van Veen, J.A. (1988) *Soil Biology and Biochemistry* **20**, 483–488.
67. Heijnen, C.E. & van Veen, J.A. (1991) *FEMS Microbiology Ecology* **85**, 73–80.
68. Heijnen, C.E., Hok-a Hin, C.H. & van Veen, J.A. (1991) *FEMS Microbiology Ecology* **85**, 65–72.
69. Helling, R.B., Kinney, T. & Adams, J. (1981) *Journal of General Microbiology* **123**, 129–141.
70. Henis, Y. & Alexander, M. (1990) *Antonie Van Leeuwenhoek* **57**, 9–20.
71. Henschke, R.B., & Schmidt, F.R.J. (1989) *Biology and Fertility of Soils* **8**, 19–24.
72. Henschke, R.B., Nucken, E. & Schmidt, F.R.J. (1989) *Biology and Fertility of Soils* **7**, 374–376.
73. Henschke, R.B. & Schmidt, F.R.J. (1990) *Current Microbiology* **20**, 105–110.
74. Herron, P.R. & Wellington, E.M.H. (1990) *Applied and Environmental Microbiology* **56** 1406–1412.
75. Herron, P.R. & Wellington, E.M.H. (1992) In *Genetic Interactions Between Microorganisms in the Natural Environment*, Eds Wellington E.M. & van Elsas, J.D. Pergamon Press, Oxford, pp. 91–103.
76. Hirsch, P.R. & Spokes, J.R. (1988) In *Risk Assessment For Deliberate Release*, Ed. Klingmüller, W. Springer-Verlag, Berlin Heidelberg, pp. 10–17.
77. Holben, W.E., Jansson, J.K., Chelm, B.K. & Tiedje, J.M. (1988) *Applied and Environmental Microbiology* **54**, 703–711.
78. Hsu, S.C. & Lockwood, J.L. (1975) *Applied Microbiology* **29**, 422–426.
79. Hynes, M.F. & O'Connell, M.P. (1990) *Canadian Journal of Microbiology* **36**, 864–869.
80. Jain, R.K., Sayler, G.S., Wilson, J.T., Houston, L. & Pacia, D. (1987) *Applied and Environmental Microbiology* **53**, 996–1002.
81. Jeffrey, W.H., Paul, J.H. & Stewart, G.J. (1990) *Microbial Ecology* **19**, 259–268.
82. Jones, I.M., Primrose, S.B., Robinson, A. & Ellwood, D.C. (1980) *Molecular and General Genetics* **180**, 579–584.

83. Kleeburger, A. & Klingmüller, W. (1980) *Molecular and General Genetics* **130**, 621–627.
84. Klein, D.A. & Casida, E. (1967) *Canadian Journal of Microbiology* **13**, 1461–1469.
85. Knudsen, G.R., Walter, M.V., Porteous, L.A., Prince, V.J., Armstrong, J.L. & Seudler, R.J. (1988) *Applied and Environmental Microbiology* **54**, 343–347.
86. Kokjohn, T.A. (1989) In *Gene Transfer in the Environment*, Eds Levy, S.B. & Miller, R.V. McGraw Hill, New York, pp. 73–97.
87. Kozyrovskaya, N.A., Gvozdyak, R.I., Muras, V.I. & Korychum, V.A. (1984) *Archives of Microbiology* **137**, 338–343.
88. Krasovsky, V.N. & Stotzky, G. (1987) *Soil Biology and Biochemistry* **19**, 631–638.
89. Kuykendall, L.L. & Weber, D.F. (1978) *Applied and Environmental Microbiology* **36**, 915–919.
90. Labeda, D.P., Kang-Chien, L. & Casida, L.E. (1976) *Applied and Environmental Microbiology* **31**, 551–561.
91. Lacy, G.H. & Leary, J.V. (1975) *Journal of General Microbiology* **88**, 49–57.
92. Levy, S.B., Marshall, B. & Rouse-Engel, D. (1980) *Science* **209**, 391–394.
93. Lewis, D.M., Bromfield, E.S.P. & Barran, L.R. (1987) *Canadian Journal of Microbiology* **33**, 343–345.
94. Liang, L.N., Sinclair, J.L., Mallory, L.M. & Alexander, M. (1982) *Applied and Environmental Microbiology* **44**, 708–714.
95. Lindow, S.E. (1987) *Applied and Environmental Microbiology* **53**, 2520–2527.
96. López-Torres, A.J., Priedo, L. & Hazen, T.C. (1988) *Microbial Ecology* **15**, 41–57.
97. Lorenz, M.G., Aardema, B.W. & Krumbein, W.E. (1981) *Marine Biology* **64**, 225–230.
98. Lorenz, M.G. & Wackernagel, W. (1987) *Applied and Environmental Microbiology* **53**, 2948–2956.
99. Lorenz, M.G., Aardema, B.W. & Wackernagel, W. (1988) *Journal of General Microbiology* **134**, 107–112.
100. Lorenz, M.G. & Wackernagel, W. (1988) In *Risk Assessment for Deliberate Release*, Ed. Klingmüller, W. Springer-Verlag, Berlin & Heidelberg, pp. 119–117.
101. Lorenz, M.G. & Wackernagel, W. (1990) *Archives of Microbiology* **154**, 380–383.
102. MacDonald, R.M. (1986) *Soil Biology and Biochemistry* **18**, 399–406.
103. Marshall, B.M., Flynn, P., Kamely, D. & Levy S.B. (1988) *Applied and Environmental Microbiology* **54**, 1776–1783.
104. Mårtensson, A.M. (1989) *Canadian Journal of Microbiology* **36**, 136–139.
105. McClure, N.C., Fry, J.C. & Weightman, A.J. (1990) In *Bacterial Genetics in Natural Environments*, Eds Fry, J.C. & Day, M.J., Chapman & Hall, London, pp. 11–129.
106. McLoughlin, T. & Dunican, L.K. (1981) *Journal of Applied Bacteriology* **50**, 65–72.
107. Meade, J., Higgins, P. & O'Gara, F. (1985) *Applied and Environmental Microbiology* **49**, 899–903.
108. Miller, H.J., Lilejeroth, E., Henken, G. & van Veen, J.A. (1989) *Canadian Journal of Microbiology* **36**, 254–258.

109. Morgan, J.A.W., Winstanley, C., Pickup, R.W., Jones, J.G. & Saunders, J.R. (1989) *Applied and Environmental Microbiology* **55**, 2537–2544.
110. Morita, R.Y. (1985) In *Bacteria in their Natural Environments*, Eds Fletcher, M. & Floodgate, G.D., Academic Press, London, pp. 111–130.
111. Morrison, W.D., Miller, R.V. & Sayler G.S. (1978) *Applied and Environmental Microbiology* **36**, 724–730.
112. Nedwell, D.B. & Gray, T.R.G. (1987) In *Ecology of Microbial Communities*, Eds Fletcher, M., Gray, T.R.G. & Jones, J.G., Society for General Microbiology Symposium No. 41. Cambridge University Press, Cambridge, pp. 21–54.
113. O'Kane, D.J., Lingle, W.L., Wampler, J.E., Legocki, M., Legocki, R.P. & Szalay, A.A. (1988) *Plant Molecular Biology* **10**, 387–399.
114. Ogram, A., Sayler, G.S. & Barkay, T. (1987) *Journal of Microbiological Methods* **7**, 57–66.
115. Onderdonk, A., Marshall, B., Cisneros, R. & Levy, S.B. (1981) *Infection and Immunity* **32**, 74–79.
116. Orvos, D.R., Lacy, G.H. & Cairns, J. (1990) *Applied and Environmental Microbiology* **56**, 1689–1694.
117. Paul, J.H. & David, A.W. (1989) *Applied and Environmental Microbiology* **55**, 1865–1869.
118. Paul, J.H., Jeffrey, W.H. & Cannon, J.P. (1990) *Applied and Environmental Microbiology* **56**, 2957–2962.
119. Pettibone, G.W., Sullivan, S.A. & Shiaris, M.P. (1987) *Applied and Environmental Microbiology* **53**, 1241–1245.
120. Pickup, R.W. (1989) *Microbial Ecology* **18**, 211–220.
121. Pillai, S.D. & Pepper, I.L. (1990) *Soil Biology and Biochemistry* **22**, 265–270.
122. Poindexter, J.S. (1981) *Advances in Microbial Ecology* **5**, 63–89.
123. Postma, J., Walter, S. & van Veen, J.A. (1989) *Soil Biology and Biochemistry* **21**, 437–442.
124. Postma, J. & van Veen, J.A. (1990) *Microbial Ecology* **19**, 149–162.
125. Postma, J., Hok-a-hin, C.H. & van Veen, J.A. (1990) *Applied and Environmental Microbiology* **56**, 495–502.
126. Powlson, D.S. & Jenkinson, D.S. (1976) *Soil Biology and Biochemistry* **8**, 179–188.
127. Pretorius-Guth, I.M., Puhler, A. & Simon, R. (1990) *Applied and Environmental Microbiology* **56**, 2354–2359.
128. Rafii, F. & Crawford, D.L. (1988) *Applied and Environmental Microbiology* **54**, 1334–1340.
129. Rafii, F. & Crawford, D.L. (1989) In *Gene Transfer in the Environment*, Eds Levy, S.B. & Miller, R.V., McGraw Hill, New York, pp. 309–345.
130. Rafii, R. & Crawford, D.L. (1989) *Current Microbiology* **19**, 115–121.
131. Ramirez, C. & Alexander, M. (1980) *Applied and Environmental Microbiology* **40**, 492–499.
132. Ramos, J.L., Duque, E. & Ramos-Gonzalez, M.I. (1991) *Applied and Environmental Microbiology* **57**, 260–266.
133. Ramsay, A.J. (1984) *Soil Biology and Biochemistry* **16**, 475–481.
134. Rattray, E.A.S., Prosser, J.I., Kilham, K. & Glover, L.A. (1990) *Applied and Environmental Microbiology* **56**, 3368–3374.
135. Recorbet, G., Givandon, A., Stienberg, C., Bally, R., Normand, P. & Faurie, G. (1992). *FEMS Microbiology Ecology* **86**, 187–194.

136. Richaume, A., Angle, J.S. & Sadowsky, M.J. (1989) *Applied and Environmental Microbiology* **55**, 1730–1734.
137. Rochelle, P.A., Day, M.J. & Fry, J.C. (1988) *Journal of General Microbiology* **134**, 2933–2941.
138. Rochelle, P.A., Fry, J.C. & Day, M.J. (1989) *Journal of General Microbiology* **135**, 409–424.
139. Rochelle, P.A., Fry, J.C. & Day, M.J. (1989) *FEMS Microbiology Ecology* **62**, 127–136.
140. Saye, D.J., Ogunseitan, O.A., Sayler, G.S. & Miller, R.V. (1987) *Applied and Environmental Microbiology* **53**, 987–995.
141. Saye, D.J., Ogunseitan, O.A., Sayler, G.S. & Miller, R.V. (1990) *Applied and Environmental Microbiology* **56**, 140–145.
142. Sayler, G.S., Fleming, J., Applegate, B., Werner, C. & Nikbakht, K. (1989) In *Recent Advances in Microbial Ecology* Eds Hattori, T., Ishida, Y., Maruyama, Y., Morita, R.Y. & Uchida, A., Japan Scientific Press, Tokyo, pp. 658–662.
143. Scanferlato, V.S., Orvos, D.R., Cairns, J. & Lacy, G.H. (1989) *Applied and Environmental Microbiology* **55**, 1477–1482.
144. Scanferlato, V.S., Lacy, G.H. & Cairns, J. (1990) *Microbial Ecology* **20**, 11–20.
145. Schauer, A.T. (1988) *Tibtech* **6**, 23–27.
146. Short, K.A., Seidler, R.J. & Olsen, R.H. (1990) *Canadian Journal of Microbiology* **36**, 821–825.
147. Sjogren, R.E. & Gibson, M.J. (1981) *Applied and Environmental Microbiology* **41**, 1331–1336.
148. Smit, E. & van Elsas, J.D. (1990) *Current Microbiology* **21**, 151–157.
149. Steffan, R.J. & Atlas, R.M. (1988) *Applied and Environmental Microbiology* **54**, 2185–2191.
150. Steffan, R.J., Goksøyr, J., Bej, A.K. & Atlas, R.M. (1988) *Applied and Environmental Microbiology* **54**, 2908–2915.
151. Stewart, G.J. (1989) In *Gene Transfer in the Environment*, Eds Levy, S.B. & Miller, R.V. McGraw Hill, New York, pp. 141–164.
152. Stewart, G.J., Carlson, C.A. & Ingraham, J.L. (1983) *Journal of Bacteriology* **156**, 30–35.
153. Stewart, G.J. & Sinigalliano, C.D. (1989) *Archives of Microbiology* **152**, 520–526.
154. Stewart, G.J. & Sinigalliano, C.D. (1990) *Applied and Environmental Microbiology* **56**, 1818–1824.
155. Stewart, G.J., Sinigalliano, C.D. & Garko, K.A. (1991) *FEMS Microbiology Ecology* **85**, 1–8.
156. Stotzky, G. (1989) In *Gene Transfer in the Environment*, Eds Levy, S.B. & Miller, R.V. McGraw Hill, New York, pp. 165–222.
157. Talbot, H.W., Yamamoto, D.Y., Smith, M.W. & Seidler, R.J. (1980) *Applied and Environmental Microbiology* **39**, 97–104.
158. Temple, K.L., Camper, A.K. & McFeters, G.A. (1980) *Applied and Environmental Microbiology* **40**, 794–797.
159. Top, E., Mergeay, M., Springael, D. & Verstaete, W. (1990) *Applied and Environmental Microbiology* **56**, 2471–2479.
160. Torsvik, V.L. (1980) *Soil Biology and Biochemistry* **12**, 15–21.
161. Torsvik, V.L. & Goksøyr, J. (1978) *Soil Biology and Biochemistry* **10**, 7–12.
162. Trevors, J.T. (1987) *Water, Air and Soil Pollution* **34**, 409–414.

163. Trevors, J.T. & Oddie, K.M. (1986) *Canadian Journal of Microbiology* **32**, 610–613.
164. Trevors, J.T. & Starodub, M.E. (1987) *Systematics and Applied Microbiology* **9**, 312–315.
165. Trevors, J.T. & Berg, G. (1989) *Systematics and Applied Microbiology* **11**, 223–227.
166. Trevors, J.T., van Elsas, J.D., van Overbeek, L.S. & Starodub, M.E. (1990) *Applied and Environmental Microbiology* **56**, 401–408.
167. Tsai, Y-L. & Olson, B.H. (1991) *Applied and Environmental Microbiology* **57**, 1070–1074.
168. Turco, R.F., Moorman, T.B. & Bezdicek, D.F. (1986) *Soil Biology and Biochemistry* **18**, 259–262.
169. Van Elsas, J.D. (1992) In *Genetic Interactions Among Microorganisms in the Natural Environment*, Eds Wellington, E.M. & van Elsas, J.D. Pergamon Press, Oxford, pp. 17–39.
170. Van Elsas, J.D., Dijkstra, A.F., Govaert, J.M. & van Veen, J.A. (1986) *FEMS Microbiology Ecology* **38**, 151–160.
171. Van Elsas, J.D., Govaert, J.M. & van Veen, J.A. (1987) *Soil Biology and Biochemistry* **19**, 639–647.
172. Van Elsas, J.D., Trevors, J.T. & Starodub, M.E. (1988) *FEMS Microbiology Ecology* **53**, 299–306.
173. Van Elsas, J.D., Trevors, J.T., Starodub, M.E. & van Overbeek, L.S. (1990) *FEMS Microbiology Ecology* **73**, 1–12.
174. Van Elsas, J.D., van Overbeek, L.S., Feldmann, A.M., Dullemans, A.M. & deLeeuw, O. (1991) *FEMS Microbiology Ecology* **85**, 53–64.
175. Van Overbeek, L.S., van Elsas, J.D., Trevors, J.T. & Starodub, M.E. (1990) *Microbial Ecology* **19**, 239–249.
176. Wang, Z., Crawford, D.L., Pometto, A.L. & Rafii, F. (1989) *Canadian Journal of Microbiology* **35**, 535–543.
177. Weinberg, S.R. & Stotzky, G. (1972) *Soil Biology and Biochemistry* **4**, 171–180.
178. Wellington, E.M.H., Saunders, V.A., Cresswell, N. & Wipat, A. (1988) In *Biology of Actinomycetes*, Eds Beppu, T. & Ogawara, H. Japan Scientific Societies Press, Tokyo, pp. 301–305.
179. Wellington, E.M.H., Cresswell, N. & Saunders, V.A. (1990) *Applied and Environmental Microbiology* **56**, 1413–1419.
180. Wellington, E.M.H., Cresswell, N., Herron, P.R., Clewlow, L.J., Saunders, V.A. & Wipat, A. (1990). In *Bacterial Genetics in Natural Environments*, Eds Fry, J.C. & Day, M.J. Chapman & Hall, London, pp. 216–230.
181. Wellington, E.M.H., Cresswell, N. & Herron, P.R. (1992) *Gene* **115**, 193–198.
182. Wessendorf, J. & Lingens, F. (1989) *Applied Microbiology and Biotechnology*, **31**, 97–102.
183. West, A.W., Burges, H.D., Dixon, T.J. & Wybon, C.H. (1985) *Soil Biology and Biochemistry* **17**, 657–665.
184. Whan Lee, S. & Edlin, G. (1985) *Gene* **39**, 173–180.
185. Williams, S.T. (1985) In *Bacteria in their Natural Environments*, Eds Fletcher, M. & Floodgate, G.D., Special Publication for the Society for General Microbiology No. 16. Academic Press, London, pp. 81–110.
186. Williams, S.T. & Davies, F.L. (1965) *Journal of General Microbiology* **38**, 251–261.

187. Winstanley, C., Morgan, J.A.W., Pickup, R.W., Jones, J.G. & Saunders, J.R. (1989) *Applied and Environmental Microbiology* **55**, 771–777.
188. Wipat, A., Wellington, E.M.H. & Saunders, V.A. (1991) *Applied and Environmental Microbiology*, **57**, 3322–3330.
189. Yeung, K.A., Schell, M.A. & Hartel, P.G. (1989) *Applied and Environmental Microbiology* **55**, 3243–3246.
190. Zeph, L.R. & Casida, L.E. (1986) *Applied and Environmental Microbiology* **52**, 819–823.
191. Zeph, L.R., Onaga, M.A. & Stotzky, G. (1988) *Applied and Environmental Microbiology* **54**, 1731–1737.
192. Zeph, L.R. & Stotzky, G. (1989) *Applied and Environmental Microbiology* **55**, 661–665.

Chapter 7

Mathematical Modelling of Genetically Engineered Microorganisms in the Environment

J.I. Prosser
Department of Molecular & Cell Biology, University of Aberdeen

INTRODUCTION

Full realization of the potential for the use of genetically engineered microorganisms (GEMs) in biotechnology requires comprehensive assessment of potential risks associated with their environmental release. To be of commercial value and fulfil its function, a GEM must affect the environment in a significant and measurable way. This may be through its influence on the indigenous flora or fauna of the environment, or by influencing the rates of biological processes. The benefits arising from the introduction of GEMs, for example elimination of a pathogen, accelerated degradation of toxic waste material, must be balanced against potential undesirable effects. Possible risks include detrimental effects on the indigenous population, detrimental effects on nutrient cycling processes or indirect effects on both populations

Monitoring Genetically Manipulated Microorganisms in the Environment. Edited by C. Edwards
Published 1993 John Wiley & Sons Ltd. © 1993 J.I. Prosser

and processes arising from transfer of genetic material to the indigenous microflora and to higher organisms.

To assess these risks, information is therefore required on the frequency of gene transfer and on the degree to which the GEM inoculum can establish within the community, i.e. its growth, activity, survival and death, and also its dispersal from the site of inoculation. In addition information on these aspects of microbial ecology is required to improve the effectiveness of GEM inocula and will be of considerable value in optimizing design and construction of efficient strains.

Detrimental effects on population structure may include the elimination of a particular species or strain or an adjustment in the species balance within the community. This may lead to dominance by indigenous organisms, previously present at low levels, or dominance by the GEM itself. Such changes, however, are only likely to be harmful if the activites of eliminated organisms are essential and unique and if those of the newly constituted population are harmful, or at least less beneficial. Similarly, dominance by the inoculum will only be a risk if its activities harm the indigenous microflora or processes.

The range of microorganisms present in natural environments and their metabolic versatility are extensive. The majority of processes involved in the microbial cycling of nutrients, and other soil processes and interactions, may be carried out by several species or genera. This significantly reduces the likelihood of abolishing any single process by eliminating any one microbial species. In addition, public sensitivity to extinction of microbial strains or species is very different to that for higher organisms. While introduction of a GEM may affect the balance within the indigenous microbial community, assessment and quantification of risks must ultimately consider effects, direct or indirect, of the inoculum on microbial processes. These may be more significant, and easier to measure, than subtle changes in species composition, and risk assessment must achieve a balance between these two aspects of the environmental impact of GEMs.

The emphasis in this chapter is on quantitative risk assessment. Most studies of the environmental impact of GEMs have been of a qualitative or semi-quantitative nature and little attempt has been made to develop or apply quantitative theoretical models to predict or explain the effects of GEM inocula. Consequently, it is only realistically possible to discuss the need for and potential advantages of modelling approaches. The different techniques and applications of mathematical models in microbial ecology will therefore be illustrated, highlighting aspects of relevance to environmental release of GEMs and identifying areas where future development is required.

MATHEMATICAL MODELLING

As indicated above, assessment of the impact of GEMs on an environment should be quantitative, implying some application of mathematics. The term mathematical modelling, however, usually represents something more complex and generally refers to the use of mathematical equations in describing a particular system or process. Within this broad definition lies a wide range of models, distinguished by the application of different modelling techniques or adoption of different approaches. For example, non-segregated models consider microbial biomass to be homogeneously dispersed throughout the system, while segregated models consider the population as a collection of individual, discrete cells. Structured models take account of, for example, internal cell or population structure, often with compartmentalization of different components, while unstructured models lump together all components. Stochastic models consider statistical variation within systems, and chance events such as mutation, and are important when considering small numbers of individuals. Predictions of deterministic models depend solely on the mathematical equations describing the system and the initial conditions.

Risk assessment of GEMs, and quantitative microbial ecology in general, are frequently concerned with changes in particular factors with time and most models consist of differential equations describing changes in, for example, population size and substrate and product concentrations. In other situations, however, we may be interested in the particular 'state' of a system and a quantitative description of this state. An example of relevance might be measurement of the distribution of individuals among different species as a species diversity index or coefficient.

A major distinction can be drawn between empirical and mechanistic modelling. The importance of this distinction lies in the relationships of theoretical models, and their predictions, to experimental studies. Empirical models are essentially descriptive and aim to identify the simplest mathematical expression capable of representing experimental data with acceptable precision. In mechanistic models, the mathematical equations represent a quantitative theory, based on simplifying environmental and biological assumptions and incorporating something new and novel, the anacalyptic assumptions (49), regarding the mechanisms controlling the system.

An empirical model arises solely and directly from experimental work. An example might be the fitting of simple linear or exponential curves to the decline in the concentration of cells following inoculation into an unfavourable environment. A wide range of techniques and experimental systems might be used to generate such data; for example, batch soil or soil slurry systems or samples of lakewater or sewage, incubated in a controlled manner but under conditions representing as closely as possible the natural environment. These are true microcosms. They aim to mimic as closely as possible a portion of the natural environment but under controlled and defined conditions, and with a level and sophistication of monitoring which would not be possible in the field. They might be used to determine the effect of environmental factors on nutrient cycling processes or microbial survival, and effects of a pollutant or introduction of a GEM. Their value lies in the simplicity with which they quantify these effects, represented as changes in the values of one or a few rate constants.

Mechanistic models, however, precede data acquisition, and experimental studies are designed to test the hypothesis being proposed by and incorporated into the model. Usually such a hypothesis will be quite specific, e.g. suggesting that the effect of substrate concentration on specific growth rate is described by the Monod equation (27), or that the rate of colonization of a surface decreases as available space decreases. It would obviously be foolish to carry out initial studies, either theoretical or experimental, on the effect of substrate concentration on specific growth rate while other environmental factors were varying. Experimental systems or microcosms used to test mechanistic models must therefore be designed to obey as closely as possible the simplifying assumptions on which the model is based. Thus, if it is implicit in the model that temperature and pH are constant, and that all cells within the system are identical, this should be achieved in the experimental systems, e.g. by maintaining axenic cultures and incubating at constant temperature with pH control. Such a system will be very different to the natural environment, where such factors will vary, but the aim is to test the anacalyptic assumptions of the model. Effects of other factors may be introduced at a later stage.

A wide range of experimental model systems have been developed to study important aspects of microbial ecology (50). It is not possible to describe fully these systems here but the study of soil nitrogen transformations will be used as an example.

Originally, transformation of nitrogen compounds was studied using simple batch incubation systems. For example, nitrification rates may be measured in samples of soil incubated with ammonium at constant temperature and moisture content, with analysis of samples for ammonium, nitrite and nitrate. Improvements in these measurements were achieved by the use of reperfusion systems (20). These consist of columns of soil through which is recycled, or reperfused, liquid containing the compounds of interest, for example, ammonium, other nutrients, or compounds such as pesticides which may influence the process. Samples taken from the reservoir are analysed for substrate and product concentrations and the system provides better control of moisture content than incubation studies and avoids disturbance of the soil when sampling. Improved control is provided by continuous flow sand or soil columns (1, 2, 26), through which the required solutions are passed continuously, with effluent samples analysed for substrate and product concentrations. These systems can be operated at controlled matric potentials (47), with pure cultures or with the indigenous populations, at constant temperature, constant or varying flow rates, etc. The systems can be further refined by incorporating glass beads, or other inert material, rather than soil to further reduce unwanted variables (3). The choice of operating conditions depends on the purpose of the study, and these systems need not necessarily be linked to a quantitative theory or model. The important point is that such experimental systems, and equivalent systems relating to aquatic environments, minimize variability in factors in which there is no immediate or direct interest. Consequently, they satisfy experimentally the simplifying assumptions of the theoretical models, enabling concentration on factors which are specifically under study.

Mathematical models, therefore, cannot be considered in isolation from practical or experimental studies. There is no single modelling technique or approach which is applicable to all situations or to solution of all problems. In assessing the impact of a GEM on an ecosystem, it is necessary initially to state which questions are being posed, to determine whether modelling and quantification are necessary and, if so, to decide on an appropriate modelling approach. It may be that very simple empirical models may suffice for certain descriptive or monitoring studies. More detailed understanding of the underlying mechanisms controlling ecosystems will require more complex mechanistic models. In many cases a combination of the two approaches may be desirable, with mechanistic models incorporating empirical components. A comprehensive treatment of all models is not possible here and the following sections provide an indication of the modelling approaches adopted in investigating aspects of microbial ecology relevant to risk assessment.

MICROBIAL GROWTH

In the presence of excess substrate and favourable conditions, microorganisms grow exponentially, as represented by the differential equation:

$$dx/dt = \mu_m x \qquad (7.1)$$

and its analytical solution:

$$x = x_0 e^{\mu_m t} \qquad (7.2)$$

where x and x_0 represent biomass concentrations at times t and 0, and μ_m is the maximum specific growth rate. This relationship may be derived from assumptions that all cells in a culture are identical, with constant and equal doubling times, and that each cell gives rise to two identical daughter cells which grow in the manner of their parent cell. The relationship also facilitates comparison of specific growth rates of different organisms, and of the effects of different factors on growth rate, by comparison of values of a single constant, μ_m. In most natural environments, however, growth is not maximal, usually being limited by availability of substrate and frequently reduced by sub-optimal environmental conditions. The effect of substrate concentration on microbial specific growth rate, μ, is most commonly described by the Monod equation (27):

$$\mu = \mu_m s/(K_s + s) \qquad (7.3)$$

where s is the concentration of the limiting substrate, K_s is the half-saturation constant for growth, equal to the substrate concentration at which $\mu = \mu_m/2$. Incorporation of the Monod equation into Equation 7.1 produces a set of two equations describing rates of change in biomass (x) and substrate (s) concentrations

$$dx/dt = \mu_m xs/(K_s + s) \tag{7.4}$$

$$ds/dt = - \mu_m xs/[Y(K_s + s)] \tag{7.5}$$

where Y is the yield coefficient for production of biomass on the limiting substrate. This model requires, implicitly, consideration of growth as conversion of substrate into biomass. As such, it is more comprehensive than Equations 7.1 or 7.2 but its increased complexity reduces usefulness for routine application. Simultaneous, analytical solution of Equations 7.4 and 7.5 is not possible and growth is now determined, or represented by three constants, μ_m, K_s and Y.

Microbial growth is also frequently modelled by the logistic equation, which may be written in the form:

$$dx/dt = \mu_m x(1 - x/x_{max}) \tag{7.6}$$

where x_{max} is the maximum biomass supportable by the system. This empirical equation is derived from similar equations describing population changes in higher organisms and involves two constants, μ_m and x_{max}. In the original derivation of the equation, μ_m is represented by r, the intrinsic rate of increase, and x_{max} by k, the carrying capacity of the environment. The equation fits growth of microbial populations in batch liquid culture and has also been applied to populations colonizing soil and other solid surfaces. Its basic mathematical assumption is that specific growth rate decreases as a negative linear function of biomass concentration or population size. This relationship was chosen as the simplest that fitted experimental data on population growth of higher organisms. The quality of fit between predicted and experimental data does not therefore, in itself, increase our understanding of microbial growth. Nevertheless, the simplicity of the model and the ability to solve Equation 7.6 analytically present significant advantages.

These three relatively simple sets of equations provide the basis for many kinetic measures of microbial growth and activity in natural environments. The simplicity of Equations 7.2 and 7.6 favour their use in quantifying the impact of a GEM, or measurement of growth of the GEM itself. The understanding provided by Equations 7.3–7.5 enables us to predict to some extent the ability of a GEM to grow in a particular environment. For example, organisms with high μ_m are likely to benefit in environments with high substrate concentrations, while those with low K_s will be better adapted to oligotrophic conditions. These factors are considered in more detail below (see later section, p. 182).

EFFECTS OF ENVIRONMENTAL FACTORS ON GROWTH

Although many environmental factors affect specific growth rate, the influence of temperature has been modelled most extensively. Until recently this was achieved by application of the Arrhenius equation, derived for the effect of temperature on chemical reactions.

$$\mu_m = Ae^{-E/RT} \tag{7.7}$$

where A is the 'collision' factor, E is the activation energy (assumed to be independent of temperature), R is the gas constant and T is the absolute temperature. Thus, μ_m is predicted to increase as an exponential function of $1/T$, up to the optimum temperature for growth, T_{opt}. Topiwala & Sinclair (42) also presented a model in which both μ_m and K_s varied in this manner and modelled the lag in adjustment to new specific growth rates following changes in temperature.

More recently, a 'square-root' model has been proposed (22, 36) and provides a much more accurate description of temperature effects in a wide range of microorganisms. This proposes a square root relationship between specific growth rate at temperatures below T_{opt} and is combined with an exponential decrease in μ_m at temperatures greater than the optimum:

$$\mu_m = b(T - T_{min})\,[1 - e^{c(T - T_{max})}] \tag{7.8}$$

where b, c, T_{min} and T_{max} are constants estimated from experimental data. The mechanistic basis for this relationship is not known but predictions of Equation 7.8 fit experimental data much more closely than the Arrhenius equation and it is increasingly being applied to the prediction of microbial spoilage of food during storage at low temperature.

Effects of other environmental factors on specific growth rate have not been modelled to the same extent but examples applicable to specific processes or organisms do exist. For example, Quinlan (34) modelled the combined effect of pH and substrate concentration on specific growth rate of ammonia-oxidizing bacteria. The substrate for ammonia oxidation is NH_3, rather than NH_4^+, and the model involved consideration of the effect of pH on the relative proportions of NH_3 and NH_4^+. Essentially, therefore, it describes the effect of pH on substrate availability and provides an example of a mechanistic approach which is successful in predicting rates of nitrification in sewage treatment plants at a range of pH values.

Theoretical and predictive models for effects of other environmental factors, for example water activity and pressure, are less well developed. In terrestrial environments, water availability will be an important factor determining both the specific growth rate and survival of microorganisms and the response to transient conditions will be particularly important as soil/water relationships will change frequently.

The effects of transient conditions generally are poorly understood, both in terms of the mechanisms by which cell physiology adjusts to changes and quantification of such changes. Studies on steady-state chemostat cultures indicate that cells are capable of responding quickly to small step changes in substrate concentration (9, 30). Significant changes, however, lead to a lag before adjustment in specific growth rate to new conditions. To an extent adjustments may be understood in terms of the time required for synthesis of ribosomes when conditions for growth become more favourable (31) and may be modelled using compartmental, structured models of the microbial cell (48). These models have not, however, been used to study transient effects in natural environments where organisms will rarely experience constant conditions for significant periods of time.

Another major factor influencing growth in both terrestrial and aquatic environments is attachment to, and colonization of solid surfaces. A combination of physico-chemical and biological mechanisms are involved in the early stages of cell attachment. Theoretical modelling of the physico-chemical mechanisms is easier, treating cells as colloidal particles, which have been the subject of extensive study (37). Biological factors, for example pilus formation and production of extracellular polymeric substances, are more difficult to describe in a quantitative fashion. This problem is increased as colonization proceeds and a mature, multilayer biofilm is formed. Conditions within such structures are heterogeneous and complex and lead to significant variation in the physiology of cells in different regions.

In addition to attachment and colonization of a surface, detachment and sloughing of cells must also be considered, again involving consideration of physico-chemical mechanisms, but complicated by the heterogeneous and poorly defined nature of microbial cells and aggregates.

Despite these difficulties, models of biofilm formation have been developed and tested in experimental model systems (5, 7). These studies are of relevance to growth and establishment of GEMs in identifying and quantifying the physiological factors important in attachment and colonization and, conversely, the effect of biofilm formation on cell physiology and activity. This understanding increases the ability to predict the behaviour of a GEM in the natural environment using a knowledge of the properties of the organism in laboratory culture. In addition, quantification of biofilm formation is important when considering other aspects of microbial ecology, for example interactions with other organisms, survival and gene transfer.

In general, therefore, mathematical models of microbial growth, and the influence of environmental factors, have been tested in laboratory experimental systems but are equally applicable to growth in natural environments. Their major role is likely to be in providing an understanding of microbial physiology in such environments, enabling prediction of the growth capabilities of GEM inocula. In addition, the simpler, more empirical models, notably those for temperature effects, are of direct predictive use, although the situation may be complicated by interactions with other factors not tested in the laboratory.

SURVIVAL

The survival of GEMs following inoculation is of importance both in terms of their effectiveness and in risk assessment. The factors affecting survival of microorganisms in natural environments are many and complex. In the absence of nutrients, microorganisms adopt a number or survival strategies, including formation of dormant spores and of vegetative cells, including ultramicrobacteria, in which metabolic activity is reduced and cells have increased resistance to stress and adverse environmental conditions (28,29,39). The physiology of starvation responses is not sufficiently well understood to enable detailed mechanistic modelling of survival and recovery of cells in natural environments. In addition, factors related to heterogeneity, e.g. surface growth, microenvironments, existence of hot spots of activity, become more important as concentrations of cells decrease, necessitating use of stochastic modelling.

Although mechanistic modelling of survival may not currently be feasible, empirical modelling may be of use. Considerable efforts have been devoted to defining the kinetics of microbial death following treatment with sterilizing agents, such as heat, irradiation and bactericidal chemicals. Typically such treatment is assumed to result in an exponential decrease in viable cell concentration with increasing time of exposure. The simplicity of this relationship has led to its widespread use but in practice the shapes of survivor curves show considerable variation, with semilogarithmic plots being linear, convex, concave, sigmoid or more complex (6). This variability can result from heterogeneity in the physiological state of cells within the population and in their environment. More significantly, variability in many cases is concerned with the nature of the particular treatment on the physiology of the cells.

A discussion of the physiological mechanisms of inactivation by sterilizing agents has limited relevance to starvation survival of GEMs in most natural environments, but it highlights a danger of using simple empirical models, also alluded to in discussing temperature effects on specific growth rate. Viable cell concentrations of microbial inocula and of viruses frequently appear to follow exponential kinetics and rate constants are used to compare effects of different factors and different organisms. While the use of such simple empirical models may be of enormous value, rigorous adherence can obscure valuable information. For example, deviation from exponential kinetics may actually indicate unsuspected heterogeneity in the population. A further problem is the accuracy of experimental data from natural environments which frequently prevents precise characterization of kinetics and distinction between predictions of contrasting models.

Mechanistic models of survival are therefore not available for application to starvation survival of inoculated GEMs, and there is a requirement both for quantitative models of physiological responses to starvation and for consideration of environmental factors, such as biofilm formation, which may protect cells from death. Meanwhile, empirical models may be valuable in quantifying survival in a descriptive manner but their use, and interpretation of values obtained, require caution.

MICROBIAL PROCESSES

Consideration of the impact of GEMs on natural environments is frequently approached at the level of the organism or the population, with concerns about persistence for long periods, elimination of indigenous species or disturbance of community structure. These effects will only be deleterious, however, if they significantly affect biological interactions within the community or reduce or alter nutrient cycling processes. In addition, the assessment of microbial numbers and biomass in natural environments is difficult. Despite the significant recent advances arising from development of molecular-based microbial detection techniques, it is still not possible to obtain sufficient reliable data for statistical analysis of the concentration of cells in natural samples with differentiation between viable, non-viable, active and non-culturable cells. In contrast, the development and availability of techniques for assessment of microbially mediated nutrient cycling processes are well advanced and facilitate measurement of process rates. This in turn facilitates testing of mathematical models of nutrient cycling, which should play an important role in assessing and predicting environmental impact of GEMs.

Process models have employed a wide range of approaches and have been applied to many natural environments. Illustrative examples presented here will concentrate on models of nitrogen transformations in the soil.

The simplest models are empirically based equations used to define the kinetics of nutrient transformations. For example, soil nitrification kinetics in incubation studies are determined from changes in product concentrations and may follow linear kinetics in short-term incubations where substrate is in excess and population levels do not change significantly. At limiting substrate concentrations, activity will follow saturation kinetics, as substrate is utilized, while long-term incubations with high initial substrate concentration and relatively low initial population levels will lead to exponential increases in product concentration. These simple kinetics serve two valuable functions. They facilitate comparison of rates under different conditions, while characterization of kinetics and deviation from these simple empirical models provides information on initial substrate and population concentrations and on the reliability of the experimental systems. For example, insufficient oxygen supply will reduce rates during incubation giving more complex kinetics. Although these deviations may be described by more complex empirical models, this both obscures deficiencies in the experimental system and reduces the value of the model, by increasing the number of constants defining process rates (32).

Empirical models of individual processes may be combined to produce more comprehensive models of nutrient cycles. Examples for both carbon and nitrogen cycles include those of van Veen *et al.* (44, 45), Jenkinson & Rayner (17) and Jenkinson & Parry (16). The last of these is concerned with transformation of soil nitrogen, which is considered in four pools: root, stubble and immobilized N; soil microbial biomass N; humus N; and inorganic N. The first three pools decay according to first-order kinetics. Root N decays to form biomass N and humus N, which in turn decay to form inorganic N. The model was fitted to experimental data in which ^{15}N-fertilizer was added to a wheat crop, with subsequent measurement of ^{15}N in the different pools over a period of several years. Predicted and experimental data agreed well and enabled estimation of fluxes through the different compartments and the turnover time of soil microbial biomass, which was found to be 1.52 y.

This type of model provides information on the rates of different processes and enables prediction of the impact of GEMs on nutrient cycling if they are known, or if they are expected to alter the rate of a particular process. Information on turnover time is also of value in indicating the specific growth rates of natural populations and the requirement for survival over long periods for successful establishment in such environments.

Compartmental models have also been used to simulate the transport of nutrients in soil, along with biological processes. An example is LEACHM (46), which considers the soil to consist of a series of horizontal segments of equal thickness. Transport of water and soluble nutrients is simulated using numerical approximation methods to solve classic water flow equations for saturated and unsaturated soils on the basis of data on soil properties. This basic model may then be modified to incorporate transport and transformations of nitrogen, pesticides and inorganic ions. Microbial transformations follow first-order kinetics and the model also includes terms for plant growth and uptake of water and nutrients.

Models of this type are designed for application to field conditions and are potentially of great value in predicting large-scale effects of GEMs whose influence on nutrient cycling is suspected. Thus, the impact of inoculation of a GEM that is known to influence the rate at which a group of organisms carry out a particular process can be simulated by modification of rate constants for that process. Such models therefore act as investigative and predictive tools.

Complex mechanistic models of microbial transformations have also been used to characterize the factors affecting nitrogen transformations in soil, paying particular attention to effects of microbial growth and biofilm formation, hydrodynamic dispersion, diffusion and ion exchange (33). These models function as hypotheses and have been tested in experimental models consisting of continuous flow soil and sand columns. They provide fundamental information on the mechanisms governing soil transformations and enable calculation of values of rate constants for cells and populations adsorbed to soil particles, rather than those obtained in liquid culture. One criticism is their concentration on either physico-chemical processes or microbial growth, with little interaction between the two approaches.

MICROBIAL INTERACTIONS

A combination of theoretical and experimental modelling has been used to elucidate the factors controlling microbial interactions. These studies provide basic information for predicting the properties of GEMs, and conditions, required for their establishment in natural environments. Most models consider interactions between two or three species in batch or continuous culture systems and most are based on equations similar to Equations 7.3–7.5, modified to describe particular interactions. For example, commensal or amensal interactions involving production by one species of substrates or inhibitors affecting a second species will require additional equations describing the rates of production of these compounds, and modification of the equations for changes in biomass concentration to account for enhancement or inhibition of growth.

Competition

As indicated earlier in this chapter, the competitive ability of an organism in the presence of nutrients depends on the concentration of the limiting substrate and the values of μ_m and K_s, relative to those of its competitors. Thus, at high nutrient concentrations organisms with high μ_m values will possess an advantage, while conditions of low substrate concentration will favour those with low K_s values. These groups can be defined respectively as zymogenous and autochthonous, following Winogradsky, or as adopting r or K strategies, following macroecologists and their application of the logistic equation to competition.

The importance of these factors in competition has been demonstrated by chemostat enrichment of cultures from seawater inocula and subsequent studies on competition in mixed cultures (14). At low, growth-limiting concentrations of lactate or phosphate, organisms were enriched with lower μ_m and K^s values than those enriched at higher concentrations obtained at high dilution rates. The substrate saturation curves of isolates enriched at low and high substrate concentrations (spirilla and pseudomonads respectively) were characterized and were found to cross over.

Mathematical modelling of competition between members of these two groups merely requires description of growth of each species by equations similar to Equation 7.3, with a common limiting substrate, and modification of Equation 7.4 to allow for utilization of substrate by both organisms. Simulation of the set of equations predicts that one species will dominate and the second will eventually be washed out. The species eliminated depends on substrate concentration and therefore on dilution rate, D, with the species with the higher μ_m dominating at high D and the species with the lower K_s dominating at low D. These predictions were confirmed experimentally in chemostats inoculated with pure cultures of *Spirillum* and *Pseudomonas*, *Pseudomonas* outcompeting *Spirillum* at high D and *Spirillum* dominating at low D. Matin & Veldkamp (24) carried out further studies on isolated strains to demonstrate the physiological properties of the two organisms leading to the differences in specific growth rates and saturation coefficients.

In fact, mathematical analysis of these equations enables indentification of the dilution rates and inflowing substrate concentrations giving exclusion of each organism and predicts a single value of D at which the different values of μ_m and K_s give rise to a steady-state substrate concentration at which specific growth rates of the two organisms are equal. Such a steady state would, however, be unstable and any variation in conditions would lead to instability and dominance by a single species. Although this analysis indicates competitive exclusion for organisms occupying the same niche and limited by the same nutrients, many factors can lead to coexistence. Most obviously, nutrient supply to the niche may vary with time, such that different species may be favoured at different times. Even in chemostat cultures, washout of the slower growing organism may take many days, by which time nutritional conditions in the natural environment would have changed.

Another major factor affecting competition is the existence of multiple substrate utilization and limitation. All organisms require, and are continually utilizing, a number of nutrients, each of which can limit growth. This has been modelled in two ways, each involving calculation of the term $s/(K_s + s)$ for each substrate. The first considers growth to be limited by the substrate for which this term has the smallest value. Growth is therefore still limited by a single substrate, but the nature of the limiting substrate is likely to vary with time as nutritional conditions change. The second, multiplicative, model represents specific growth rate as the product of μ_m and each $s/(K_s + s)$ term.

Taylor and Williams (41) analysed models of this type and predicted that the maximum number of species which could coexist was one less than the number of substrates limiting growth. While this conclusion is supported by some experimental work, Gottschal *et al.* (13) found coexistence of three species when only two substrates were limiting growth. They investigated competition between an obligately autotrophic sulphur oxidizer, *Thiobacillus neapolitanus*, a facultative heterotroph, *Thiobacillus* A2 (now designated *Thiobacillus versutus*) and an obligately heterotrophic *Spirillum* species. Chemostats inoculated with all three species were supplied with thiosulphate and acetate in varying proportions. At the higher thiosulphate concentrations the obligate autotroph *T. neapolitanus* dominated in the steady state, while *Spirillum* dominated at high levels of acetate. At intermediate levels the facultative species was also present, and at roughly equal propotions of thiosulphate and acetate all three organisms coexisted. Thus, three organisms coexisted on two limiting nutrients. This indicates that either the theoretical model of Taylor & Williams is incorrect or that other factors, not considered by the model, were operating.

Commensalism and inhibition

Commensalism can also lead to coexistence and has been modelled by Megee *et al.* (25) for the interaction between *Lactobacillus plantarum*, growing in a glucose-limited chemostat and producing lactate which is utilized by *Propionibacterium shermanii* under anaerobic conditions. Stability analysis of the model predicted combinations of dilution rate and inflowing substrate concentration leading to coexistence of both organisms, elimination of *P. shermanii* and elimination of both organisms.

Inhibition can relieve competition and lead to coexistence of organisms with similar nutritional requirements. de Freitas & Fredrickson (10) analysed chemostat equations describing growth of two species on a common limiting nutrient and formation of inhibitors active against the producer organism or the competitor. Production of autoinhibitors resulted in a degree of self-regulation of growth, reducing competition for the limiting nutrient, and enabled stable coexistence of both populations over a range of dilution rates and inflowing substrate concentrations. Extension of the analysis to multi-species systems indicated that production of an autoinhibitor which was common to all species competing for the limiting nutrient would lead to coexistence of two species only, specific growth rate being controlled by two compounds. If populations produced autoinhibitors specific to themselves, however, the potential for coexistence was increased and their analysis indicated that the number of species that can coexist in the steady state equals the number of specific autoinhibitors plus one. In addition, they demonstrated that production of compounds capable of inhibiting competitors can result in dominance of producer species under conditions which would otherwise be unfavourable.

Predation

Predation by protozoa and soil animals is likely to be a major factor in the survival and establishment of GEM microbial inocula and has attracted a number of modelling approaches. Early work on microbial predation involved application of the Lotka–Volterra equations, derived to describe predation in higher organisms. These equations are based on the logistic equation, with prey growth rate limited by population size, while predator growth rate is controlled by prey numbers. These equations predict oscillations in prey and predator numbers, with time, but are not capable of describing the variety of dynamics observed in microbial systems. In particular, perturbations result in instability and the equations do not describe environmental control of oscillations, which appears to be more important (38). This led to formulation of more complex models based on substrate saturation kinetics, equivalent to the Monod equation, for both prey and predator. These models are less empirical in nature and predict oscillations in both populations, damped oscillations and monotonic approaches to steady states (49). Stability analysis of the equations (43, 49) enables identification of the properties of the organisms necessary for coexistence or extinction of one or both populations, and whether oscillations will be of constant amplitude, damped or absent. In addition, the effects of increasing nutrient supply and heterogeneity in the environment can be introduced. Williams (49) discusses the ability of such models to describe data obtained from experimental model systems, most of which involve study of predation of bacteria by protozoa or slime molds. Interestingly, the model predicts behaviour in batch culture better than that in continuous culture.

Catastrophe theory has also been applied to microbial predation (4). This approach considers sudden changes in the dynamic state of a system, as opposed to the smooth changes predicted by differential equations. The theory was used to explain predation of *Escherichia coli* by the slime mold. *Dictyostelium discoideum* in continuous culture (11). Oscillations were observed in prey and predator populations but could not be described by the Lotka–Volterra equations. The specific rate of change in predator numbers remained constant for long periods, during which prey numbers varied considerably, and then changed abruptly to a new rate of change which was maintained until the next abrupt change. Predator specific growth rate is usually considered to be dependent on prey numbers, but application of catastrophe theory taking into account this relationship could not explain the observed data. When the analysis was repeated with predator specific growth rate controlled by the ratio of prey to predator numbers a good fit was obtained. The study therefore increased our understanding of the factors controlling the interaction, indicating this to be a compound produced by the prey which was then modified by the predator before exerting its effect.

The above discussion demonstrates a common approach to the use of theoretical models in the study of a range of microbial interactions. Most are mechanistic models, tested experimentally in well-defined and controlled experimental systems and are not directly applicable to interactions in natural environments. Their role is in increasing our understanding of factors controlling microbial interactions and the properties of an organism which are likely to increase its competitive ability, and enhance its survival.

SPECIES DIVERSITY AND STABILITY

A perceived and potential problem arising from the introduction of GEMs into the environment is changes in species diversity and community composition; this has generally been treated in a qualitative manner. To an extent this reflects difficulties in classifying and identifying microorganisms, a lack of uniformity in the efficiency of extraction and enrichment of different microbial groups and physiological states, and inadequate means of expressing data quantitatively.

Traditionally, quantification of species diversity and abundance has been achieved using cultural techniques for enrichment and isolation of organisms on laboratory media (18, 23, 35). Pure cultures are then subjected to a wide range of morphological, biochemical and physiological tests and grouped using numerical taxonomic techniques. A level of similarity is then defined for distinction of species and the number of isolates belonging to each species determined. Finally, a species diversity index or coefficient is calculated which aims to describe both the number of species present (abundance) and the distribution of individuals between species (evenness). Major disadvantages in this approach are difficulties in extracting all cells from the environment with equal efficiency, subsequent culturing of all extracted cells on laboratory media and ignorance of the contribution of non-culturable cells. While modern molecular-based techniques, involving nucleic acid extraction, are potentially capable of solving some of these problems, their widespread use in determining species diversity indices is not yet feasible.

The effects on microbial processes resulting from changes in numbers of organisms carrying out specific processes were considered above. An alternative approach is to consider, in more general terms, the interactions between all members of a community and effects of changes in community composition on stability of ecosystems. This approach has been used to analyse communities of higher organisms and is based on modification of the logistic equation to incorporate positive, negative or neutral interactions between all species (12). This generates a system of equations, whose stability properties can be analysed for communities containing different numbers of species and with differences in the number and type of interaction. An unexpected finding of this analysis was that stability increased with decreasing number of species and with fewer interactions.

This approach to the stability of multispecies ecosystems, and other related studies, have been developed for analysis of populations of higher organisms but have not been applied to microorganisms. Nevertheless they provide a basis for theoretical studies of the stability of microbial ecosystems, which may be more amenable to experimental analysis and testing than higher organisms.

MICROBIAL MOVEMENT AND DISPERSAL

Microorganisms may be dispersed via a wide range of mechanisms in aquatic and terrestrial environments, and in the atmosphere. This topic has recently been reviewed extensively (15) and will be dealt with only briefly here.

Dispersal of microorganisms in the atmosphere through aerosols has been studied due to the need to quantify the spread of animal and plant pathogens. In aquatic and terrestrial environments, the situation is complex because of their greater heterogeneity and the large variety of dispersal mechanisms. For example, in the soil, models have been based on those describing water flow, discussed in an earlier section (p. 181), with further consideration of factors such as the force needed to remove organisms from surfaces and the influence of cell surface properties. Other factors, e.g. the role of soil animals and the existence of microenvironments and hot spots of activity, are more difficult to incorporate. Nevertheless, many of these models have been developed for risk assessment of non-genetically engineered organisms, e.g. pathogens, and consequently are directly applicable to risk assessment of GEMs.

GENE TRANSFER

The construction of GEMs with genetic material introduced by recombinant DNA technology necessitates consideration not only of survival of the introduced organisms, but also persistence of the introduced genetic material. While it may be possible to control and limit the spread of an inoculum or choose one that is unlikely to grow and establish successfully, this will be of limited value if genetic material is transferred to members of the indigenous microbial population.

Gene transfer is important both for risk assessment and for the efficiency of the GEM within the environment. In itself, it does not constitute a risk and any danger lies in the nature of the genetic material being transferred, and the effect it will have on recipient strains and subsequently on the environment. Any quantification of risk through gene transfer must therefore consider not only the frequency of gene transfer but also the magnitude and probability of potential harmful effects. In addition commercial viability of a GEM may be compromised by gene transfer to the indigenous microflora, in which it may persist for longer periods, eliminating the need for further inoculation.

Gene transfer in the environment has been extensively studied in the past to determine the factors affecting the spread of antibiotic resistance through plasmid transfer, particularly in hospitals. Only recently have these studies been extended to gene transfer in environments such as soil, water and sediments. Transfer of genetic material is known to take place between the major microbial groups and between microorganisms and plants and there is no reason to think that gene transfer processes occurring in the laboratory do not take place in natural environments. The challenge is, therefore, to quantify such processes and to determine the environmental factors affecting transfer. In addition, gene transfer is an important ecological process, enabling the spread of properties and characteristics throughout the microbial population.

Consequently, the past few years have seen the development of genetic and experimental systems to study gene transfer. Attention has focused on conjugation, with some work on transduction and transformation with naked DNA. These studies have provided quantitative estimates of gene transfer frequencies but very few have involved construction or testing of mathematical models for gene transfer, and few of these have dealt specifically with the factors affecting gene transfer in natural environments.

Gene transfer may occur by conjugation, transduction or transformation of free or naked DNA. The first of these requires contact between donor and recipient cells mediated by production of sex pili by the donor which react with specific sites on the recipient. Gene transfer by transduction is mediated by phage. In generalized transduction, genes from the bacterial host are transferred randomly through packaging of disintegrating chromosomal DNA in the phage capsid. Specialized transduction involves lysogenic phage, which integrate with the host chromosome, transferring chromosomal genes adjacent to the site of integration. Transformation involves uptake of naked DNA by competent cells and integration of DNA into the genome.

General considerations

Modelling gene transfer in natural environments requires consideration of two basic factors: the probability, rate or frequency with which donor cells or phage, or naked DNA, come into contact with a recipient cell, and the probability, rate or frequency of gene transfer following this contact. Together, these encompass all aspects of microbial ecology and microbial physiology and the list of factors requiring quantification or comprehensive modelling is therefore effectively limitless. Predicting the frequency of contact requires consideration of all factors determining cell concentration, cell movement and cell separation. These include the concentration of the inoculum, its form and the conditions prevailing when introduced, growth and death of the host, recipient cells and phage, degradation of DNA, predation, movement by animals, plants and physical forces, spatial organization of cells, biofilm formation and establishment of hot spots where cells are concentrated, e.g. the rhizosphere.

The factors determining the probability of gene transfer following contact between bacterial cells and other cells, phage or free DNA are poorly characterized, and are rarely considered in environmental gene transfer studies. For example, information is required on the quantitative effects on gene transfer of the growth rate, activity and physiology of donor and recipient cells, environmental control of expression of genes required for plasmid transfer, the factors which make cells competent for transformation and effects of changes in cell surface properties and production of extracellular polymeric material.

The above lists are not exhaustive, and effectively demonstrate the need to know everything about microbial physiology and about the physics, chemistry and biology of the environment under study, to provide a comprehensive mathematical model of gene transfer. Obviously this is neither practical nor possible. We must therefore ask what function modelling can have in the study of gene transfer. As always, this depends on the questions being asked. For some studies an approximate 'ball-park' gene transfer frequency value may be all that is required, in which case modelling may not be necessary. If, however, information is required on the effect of a specific environmental factor on transfer frequency, mechanistic modelling, coupled to experimental testing of model predictions is the only realistic approach available.

Existing models have not achieved this, considering only very basic aspects of gene transfer and all have concentrated on plasmid transfer, considering gene transfer only in term of the probability of cells meeting. These models are based on those for gene transfer in liquid culture (21) and have the structure outlined below.

Models of gene transfer

In the absence of growth, changes in numbers of donor, recipient and transconjugant cells may be represented by the following set of equations:

$$D_t = D_{t-1} \tag{7.9}$$

$$T_t = T_{t-1} + \tau[(D_{t-1} + T_{t-1})\, R_{t-1}] \tag{7.10}$$

$$R_t = R_{t-1} - \tau[(D_{t-1} + T_{t-1})\, R_{t-1}] \tag{7.11}$$

where D, R and T represent numbers of donor, recipient and transconjugants at time intervals t and $t - 1$, represented by subscripts, and τ is the gene transfer frequency constant. The model assumes cells to be non-growing but active and homogeneously mixed, so that the number of transformants equals the number at the previous time interval, plus the number created by contact between donors, or transconjugants, and recipients. This will be proportional to the numbers, or concentrations, of the three different cell types. The transfer frequency constant, τ, quantifies the probability of cells of different types (donors and recipients) coming into contact. It is presumably related in some way to the rate at which cells move and the ratio of the volume of an individual cell to the total culture volume. In these models, the probability of gene transfer following cell contact is assumed to be unity or is a component of the transfer frequency constant. This assumption appears to be implicit and is rarely, if ever, stated.

If all cells are growing at the same rate, μ, the equations are modified to give:

$$D_t = D_{t-1} e^{\mu dt} \tag{7.12}$$

$$T_t = T_{t-1} e^{\mu dt} + \tau[(D_{t-1} + T_{t-1})\, R_{t-1}] \tag{7.13}$$

$$R_t = R_{t-1} e^{\mu dt} + \tau[(D_{t-1} + T_{t-1})\, R_{t-1}] \tag{7.14}$$

where $dt = t - (t-1)$. In many natural environments, however, cells will be dying rather than growing exponentially. To model this Knudsen *et al.* (19) incorporated a specific 'growth-death' rate (β), which could be positive or negative, depending on whether the organism was growing or dying, respectively. Using logarithms to base 10, rather than base e, assuming a time interval of 1 h and allowing for differences between growth or survival of cells with and without the plasmid gives the equations:

$$D_t = D_{t-1} \times 10^{\beta_1} \tag{7.15}$$

$$T_t = T_{t-1} \times 10^{\beta_1} + \tau[D_{t-1} + T_{t-1})\, R_{t-1}] \tag{7.16}$$

$$R_t = R_{t-1} \times 10^{\beta_2} + \tau[D_{t-1} + T_{t-1})\, R_{t-1}] \tag{7.17}$$

Here, β_1 and β_2 represent the specific death/growth rates of donor and recipient cells respectively. Knudsen *et al.* (19) tested the model with data on conjugation between two pseudomonads, one carrying a transmissible plasmid, encoding antibiotic resistance. Experiments were carried out in rhizosphere and phyllosphere microcosms. The former consisted of a peat–vermiculite mixture (PVM) in pots seeded with radish seeds and with donor and recipients, initially at 10^8 cells per g soil. Pots were then incubated in light (28–30°C) and dark (22°C) cycles for 12 d. For phylloplane studies,

3-week-old radish or bean plants were sprayed with donor and recipient cell suspensions (5×10^8 ml^{-1}) and incubated for 14 d at two different relative humidities.

For each microcosm survival rates (β) were determined for donor and recipient alone and transfer rates were estimated in sterile PVM and on leaf surfaces over 1 h. These values were then incorporated into Equations 7.15–7.17 and predictions generated for the appearance of transconjugants. Similarities between predicted and experimental results were close, with an initial increase in transconjugant numbers followed by a decline in all cases. Numbers of all populations were generally within 1 log value of those predicted with survival rates slightly better than predicted and bean survival data predicted better than for radish.

Gene transfer between streptomycetes in soil has also been modelled (8) using a modified form of the logistic equation to describe nutrient-limited growth. The model was tested in sterile and non-sterile soil microcosms, with growth parameters determined in the former. The model provided a good description of experimental data on numbers of donor, recipient and trans-conjugant populations.

Although describing experimental data well these models have many limitations, in particular lack of consideration of factors specific to soil, rather than liquid culture, and were only tested at high cell concentrations. Experiments on agar surfaces (40) have shown that, whereas plasmid transfer rates at high cell concentrations are similar to those in liquid culture, at low cell concentrations larger intercellular distances greatly reduce transfer.

Models do not yet exist which take into account the spatial organization and separation of cells and factors controlling their movement. Surfaces are of obvious importance in soil but even in aquatic environments, microbial populations will be concentrated in biofilms, potentially increasing the frequencies for gene transfer. Quantitative, model-based studies are therefore required on the factors controlling the spatial separation of cells, the abundance of 'hot spots' where cells congregate, the significance of hot spots in relation to 'dispersed' cells and factors controlling cell movement. Such studies would be equally important for modelling of gene tranfer by conjugation, transduction or transformation, although the last two would also require knowledge of movement/survival of phage and DNA.

A second important area requiring study is consideration of the effects of environmental and physiological factors on the ability of a cell to donate or receive genetic material. This will require testing of mechanistic models in well-defined laboratory culture systems, followed by microcosm work, but the major problems are likely to involve assessment of the physiological state of cells in the soil. A third, related area is the ability of 'established' populations to transfer genetic information. Most gene transfer studies have involved relatively recently inoculated populations, with the exception of those assessing transfer to the indigenous population. When cells are inoculated into natural environments, extractable, viable cell concentrations usually decrease to a relatively constant level. The factors determining this level are poorly characterized, and the potential of such populations to transfer genetic information is unknown. For example, it is not known whether ultramicrobacteria or cells adopting other survival strategies are capable of gene transfer.

SUMMARY AND FUTURE PROSPECTS

Mathematical modelling has not yet been applied extensively to assessment of the risks associated with environmental relase of GEMs, or to improvement in their efficiency. This chapter has consequently summarized the range of existing models, and modelling approaches, used to describe the ecology of microorganisms of relevance to GEMs.

Models serve several functions. They facilitate measurement of microbial growth and activity, and their influence on nutrient cycling processes, particularly in microcosms. Mechanistic models can act as quantitative theories which, in conjunction with experimental testing, can increase our understanding of factors controlling microbial ecosystems. A model which has been verified experimentally can then be used as an investigative tool, enabling assessment of effects of changes in model components of ecological relevance. Finally, some models are capable of predicting ecosystem behaviour in the field.

As with much of the debate regarding use of GEM inocula, extensive information is already available from studies on non-genetically engineered organisms. For example, models for microbial dispersal or for measurement of rate processes exist and will be equally valid for engineered and non-engineered organisms. Indeed, some of these models have been designed specifically for purposes equivalent to risk assessment, e.g. spread of pathogens. In many cases, therefore, the only requirement is for application of appropriate existing models.

The major area for application of new models is in quantifying gene transfer in natural environments. This has not been tackled in depth in the past and existing models are based on gene transfer in liquid culture with homogeneous populations of cells. In natural environments the situation is more complex and the limited experimental evidence available suggests that factors such as surface attachment and biofilm formation will significantly affect gene transfer frequencies. In other areas, there is a need for more interaction with, and application of approaches and techniques used to model the ecology of higher organisms. This applies particularly to studies on species diversity and the stability of multispecies ecosystems. Studies in these areas have been hampered by experimental and practical difficulties but realization of the potential offered by molecular-based techniques for the quantification and identification of microorganisms in natural environments should facilitate such studies and increase the potential for testing predictions of theoretical models.

The need for a modelling approach is two-fold. Firstly, a full understanding of a system requires the ability to describe and explain that system in a quantitative manner, rather than qualitatively. Modelling is therefore necessary if we are to fully understand the mechanisms controlling microbial growth, activity, dispersal and interactions in natural ecosystems. Secondly, realistic judgement of the commercial use of GEMs requires a comparison of the benefits which they provide and the risks which they pose. If this process is to be valid, both benefits and risks must be assessed quantitatively, making some form of mathematical modelling essential.

REFERENCES

1. Ardakani, M.S., Rehbock, J.T. & McLaren, A.D. (1973) *Soil Science Society of America Proceedings* **31**, 53–56.
2. Ardakani, M.S., Rehbock, J.T. & McLaren, A.D. (1974) *Soil Science Society of America Proceedings* **38**, 96–99.
3. Bazin, M.J. & Saunders, P.T. (1973) *Soil Biology and Biochemistry* **5**, 531–543.
4. Bazin, M.J. & Saunders, P.T. (1978) *Nature* **275**, 52–54.
5. Bryers, J.D. (1988) In *Physiological Models in Microbiology*, Eds Bazin, M.J. & Prosser, J.I. CRC Press, Boca Raton, Florida, pp. 109–144.
6. Casolari, A. (1988) In *Physiological Models in Microbiology, Vol. 2*, Eds Bazin, M.J. & Prosser, J.I. CRC Press, Boca Raton, Florida, pp. 1–44.
7. Characklis, W. & Wilderer, P. (Eds) (1989) *Structure and Function of Biofilms*. John Wiley & Sons, Chichester.
8. Clewlow, L.J., Cresswell, N. & Wellington, E.M.H. (1990) *Applied and Environmental Microbiology* **56**, 3139–3145.
9. Cunningham, A. & Nisbet, R.M. (1983) In *Mathematical Models in Microbiology*, Ed. Bazin, M.J. Academic Press, London, pp. 77–103.
10. de Freitas, M.J. & Fredrickson, A.G. (1978) *Journal of General Microbiology* **106**, 307–320.
11. Dent, V.A.E., Bazin, M.J. & Saunders, P.T. (1976) *Archives of Microbiology* **109**, 187–194.
12. Gardner, M.R. & Ashby, W.R. (1970) *Nature* **228**, 284.
13. Gottschal, J.C., Vries, de S. & Kuenen, J.G. (1979) *Archives of Microbiology* **121**, 241–249.
14. Harder, W., Kuenen, J.G. & Matin, A. (1977) *Journal of Applied Bacteriology* **43**, 1–24.
15. Hurst, C.J. (Ed.) (1991) *Modelling the Environmental Fate of Microorganisms*. American Society of Microbiology, Washington D.C.
16. Jenkinson, D.S. & Parry, L.C. (1989) *Soil Biology and Biochemistry* **21**, 535–541.
17. Jenkinson, D.S. & Rayner, J.H. (1977) *Soil Science* **123**, 298–305.
18. Kaneko, T., Atlas, R.M. & Krichevsky, M. (1977) *Nature* **270**, 596–599.
19. Knudsen, G.R., Walter, M.V., Porteous, L.A., Prince, V.J., Armstrong, J.L. & Seidler, R.J. (1988) *Applied and Environmental Microbiology* **54**, 343–347.
20. Lees, H. & Quastel, J.H. (1946) *Biochemical Journal* **40**, 803–812.
21. Levin, B.R., Stewart, F.M. & Rice, V.A. (1979) *Plasmid* **2**, 247–260.
22. McMeekin, T.A., Olley, J. & Ratkowsky, D.A. (1988) In *Physiological Models in Microbiology, Vol. 1.* Eds Bazin, M.J. & Prosser, J.I., CRC Press, Boca Raton, Florida, pp. 75–89.

23. Martin, Y.P. & Bianchi, M.A. (1980) *Microbial Ecology* **5**, 265–279.
24. Matin, A. & Veldkamp, H. (1978) *Journal of General Microbiology* **105**, 187–197.
25. Megee, R.D., Drake, J.F., Fredrickson, A.G. & Tsuchiya, H.M. (1972) *Canadian Journal of Microbiology* **18**, 1733–1742.
26. Misra, C., Nielsen, D.R. & Biggar, J.W. (1974) *Soil Science Society of America Proceedings* **28**, 294–299.
27. Monod, J. (1942) *Recherches sur la Croissance des Cultures Bacteriennes*, 2nd edn. Hermann, Paris.
28. Morita, R.Y. (1985) In *Bacteria in Their Natural Environments*, Eds Fletcher, M. & Floodgate G. Academic Press, New York, pp. 111–130.
29. Morita, R.Y. (1988) *Canadian Journal of Microbiology* **34**, 436–441.
30. Pickett, A.M. (1982) In *Microbial Population Dynamics*, Ed. Bazin, M.J. CRC Press, Boca Raton, Florida, pp. 1–44.
31. Pickett, A.M., Bazin, M.J. & Topiwala, H.H. (1979) *Biotechnology and Bioengineering* **15**, 1043–1055.
32. Prosser, J.I. (1988) *Advances in Microbial Ecology* **11**, 263–304.
33. Prosser, J.I. & Bazin, M.J. (1988) In *A Handbook of Laboratory Systems for Microbial Ecosystem Research*, Ed. Wimpenny, J.W.T. CRC Press, Boca Raton, Florida, pp. 31–49.
34. Quinlan, A.V. (1984) *Water Research* **18**, 561–566.
35. Rashit, E. & Bazin, M.J. (1987) *Microbial Ecology* **14**, 101–112.
36. Ratkowsky, D.A., Lowry, R.K., McMeekin, T.A., Stokes, A.N. & Chandler, R.E. (1983) *Journal of Bacteriology* **154**, 1222–1226.
37. Rutter, P.R. & Vincent, B. (1988) In *Physiological Models in Microbiology, Vol. 2*, Eds Bazin, M.J. & Prosser, J.I. CRC Press, Boca Raton, Florida, pp. 88–107.
38. Saunders, P.T. (1983) In *Mathematical Models in Microbiology*, Ed. Bazin, M.J. Academic Press, London, pp. 105–138.
39. Siegele, D.A. & Kolter, R. (1992) *Journal of Bacteriology* **174**, 345–348.
40. Simonsen, L. (1990) *Journal of General Microbiology* **136**, 1001–1007.
41. Taylor, P.A. & Williams, P.J.leB. (1975) *Canadian Journal of Microbiology* **21**, 90–98.
42. Topiwala, H.H. & Sinclair, C.G. (1971) *Biotechnology and Bioengineering* **13**, 795–813.
43. Tsuchiya, H.M., Drake, J.F., Jost, J.L. & Fredrickson, A.G. (1972) *Journal of Bacteriology* **110**, 1147–1153.
44. van Veen, J.A., Ladd, J.N. & Amato, M. (1985) *Soil Biology and Biochemistry* **17**, 747–756.
45. van Veen, J.A., Ladd, J.N. & Frissel, M.J. (1984) *Plant and Soil* **76**, 257–274.
46. Wagenet, R.J. & Hutson, J.L. (1987) In *Continuum, Vol. 2*. Water Resources Institute, Cornell University, Ithaca, New York.
47. Wagenet, R.J., Biggar, J.W. & Nielsen, D.R. (1977) *Soil Science Society of America Proceedings* **41**, 896–902.
48. Williams, F.M. (1967) *Journal of Theoretical Biology* **15**, 190-207.
49. Williams, F.M. (1980) In *Contemporary Microbial Ecology*, Eds Ellwood, D.C., Hedger, J.N., Latham, M.J., Lynch, J.M. & Slater, J.H. Academic Press, London, pp. 349–375.
50. Wimpenny, J.W.T. (Ed.) (1988) *A Handbook of Laboratory Systems for Microbial Ecosystem Research*, CRC Press, Boca Raton, Florida.

Index